Walter Miles and His 1920 Grand Tour of
European Physiology and Psychology Laboratories

THE CENTER FOR THE HISTORY OF PSYCHOLOGY SERIES

The Center for the History of Psychology Series
David B. Baker, Editor

C. James Goodwin and Lizette Royer, Editors, *Walter Miles and His 1920 Grand Tour of European Physiology and Psychology Laboratories*

Walter Miles and His 1920 Grand Tour of European Physiology and Psychology Laboratories

Walter Miles

Edited by C. James Goodwin and Lizette Royer

A REPRODUCTION OF THE ORIGINAL TYPESCRIPT,
WITH A FOREWORD BY DAVID B. BAKER AND
AN INTRODUCTION BY C. JAMES GOODWIN

The University of Akron Press
Akron, Ohio

Note to Readers

In order to print a quality reproduction of the Walter Miles report, the typescript itself was scanned in its entirety. The scans were then painstakingly cleaned to increase the contrast in order to create a print-ready file. A compact disc that contains the original scans showing the typescript prior to any adjustments is available free of charge to book buyers. In order to receive a copy, please email the following information to uapress@uakron.edu: your name, place of purchase, and mailing address. Please put "Miles CD" in the subject line of the email. Thank you.

All New Material Copyright © 2010 by The University of Akron Press
All rights reserved • First Edition 2010 • Manufactured in the United States of America. • All inquiries and permission requests should be addressed to the Publisher, the University of Akron Press, Akron, Ohio 44325-1703.
14 13 12 11 10   5 4 3 2 1
ISBN 978-1-931968-85-0
The paper used in this publication meets the minimum requirements of American National Standard for Information Sciences—Permanence of Paper for Printed Library Materials, ANSI z39.48–1984. ∞

Cover photo and backcover detail (original photo pictured above): Archives of the History of American Psychology, The University of Akron, Walter R. and Catharine Cox Miles papers. L to R: Thorne M. Carpenter; Walter B. Cannon; Walter R. Miles; H. J. Hamburger; Francis G. Benedict. The image has been tinted, cropped, and repairs have been made for use on the cover. Cover design: Amy Freels

Walter Miles and His 1920 Grand Tour was designed and typeset by Amy Freels. The text was set in Goudy Oldstyle. *Walter Miles and His 1920 Grand Tour* was printed on 60# Natural and bound by BookMasters of Ashland, Ohio.

Contents

Foreword	*David B. Baker*	vii
Introduction	*C. James Goodwin*	ix
Editorial Note		xxi

Walter Miles and His 1920 Grand Tour of European
   Physiology and Psychology Laboratories   1

    Great Britain   7
    France   127
    Belgium   167
    The Netherlands   191
    Denmark   239
    Sweden   259
    Germany and Austria   275
    General Remarks   297

Transcriptions   311
Outline   321
Biographical Profiles   337
Index   355

Foreword
David B. Baker

It is a pleasure to introduce the inaugural volume in the Center for the History of Psychology Series. This series is the realization of a long-held goal to make the historical treasure located within the Archives of the History of American Psychology more accessible. The timing could not be better. In its forty-fifth year of operation, the Archives of the History of American Psychology is undergoing dramatic growth and expansion. The publication of this volume coincides with the opening of the new Center for the History of Psychology at The University of Akron. The center provides expanded space for the Archives of the History of American Psychology, a museum, and ample room for expansion. With these changes comes a greater opportunity to educate the public about the science and practice of psychology, while maintaining the core mission of promoting research in the history of psychology.

The Center for the History of Psychology Series is designed to highlight unpublished works of historical significance and interest. The first offering was an easy choice. The Walter R. and Catharine Cox Miles Papers were one of the earliest acquisitions of the Archives of the History of American Psychology and they certainly have stood the test of time. The collection, which includes more than 43 linear feet of material, is a tour de force in the history of American psychology. From the laboratory to the field, Walter and Catharine Cox Miles sampled most of the terrain of twentieth century psychology in America. Walter Miles was fastidious in his observations and had the good sense to safely keep and protect his own archival record.

This facsimile reproduction of his 1920 trip to Europe offers a rare firsthand account of the work of early twentieth century physiology and psychology laboratories. Reading over the observations and reflections, which are richly supplemented with photographs, it is easy to transport back in time and join Miles on his visit 90 years ago. The details are rich, the personal insights and interactions revealing, and the value to the historical record priceless. There is something here for everyone interested in the history of human science.

Editors C. James Goodwin and Lizette Royer have done wonderful work providing the "bookends" for this facsimile. Their work provides important context that makes this volume readily accessible. A special thank you to AHAP student assistant Andrew White for scanning the manuscript. Acknowledgments are also due to the University of Akron Press. Their enthusiasm and support for this series is deeply appreciated.

Introduction
C. James Goodwin

In April 1920, the American experimental psychologist Walter R. Miles, then a young research scientist at the Carnegie Nutrition Laboratory in Boston, embarked on a remarkable four and a half month voyage to Europe. His goal was to visit laboratories of physiology and psychology in Great Britain and on the continent, reestablishing links that had been fractured during World War I. Between April 1920 and his return the following August, Miles visited 57 laboratories and institutes, and 9 different countries (one of them, Switzerland, was for a three-day holiday just before his return to Boston). He also attended and made presentations on apparatus of his invention to a meeting of the British Psychological Association at Cambridge and to an international conference for physiologists in Paris. Miles kept extensive records during his grand tour, converting them into a highly detailed *Report of a Visit to Foreign Laboratories: April to August, 1920*, now part of the Walter R. and Catharine Cox Miles collection in The Archives of the History of American Psychology at The University of Akron. In addition to the narrative, Miles included in his report dozens of photos and postcards accumulated during the trip. Including pages with photos or postcards attached, the report runs to more than 300 pages.

The Miles report paints a detailed and telling portrait of the state of European physiology and psychology in the years immediately after World War I. This monograph reproduces that report in facsimile form. In addition, the narrative is accompanied by a detailed outline summarizing the tour, and brief biographical portraits of the major physiologists and psychologists visited by Miles.

Walter R. Miles (1885–1978): A "Lab Man"

Walter Miles was the prototypical psychological scientist. He made important contributions to a variety of research areas in experimental psychology, was skilled as an apparatus inventor, and was happiest when immersed in the daily minutiae of laboratory life. His scientific interests were eclectic, ranging from basic research on maze learning to applied research on the physiological and psychological effects of alcohol consumption, and his projects often followed directly from his fascination with research apparatus (Goodwin, 2003). His lifelong devotion to the laboratory was recognized by a colleague who wrote that Miles was "one of the few psychologists who started out as an experimentalist and continued his interest in scientific problems all during his career, stopping neither to become a philosopher or a stamp collector or a huckster of psychological wares" (Helson, 1953). In the jargon of the day, Miles was widely recognized as a "lab man."

Miles earned his doctorate in psychology in 1913 at the University of Iowa, a student of Carl Seashore, who was well-known for his research on auditory perception and the experimental psychology of music. Miles's dissertation, on voice accuracy during pitch singing, bore the Seashore stamp. The Iowa experience also nurtured in Miles his love of research apparatus—in an obituary, Ernest Hilgard (1980) noted that Miles was "delighted with the tonoscope that Seashore had invented, and his fascination with gadgetry ... was now fixed for life" (p. 565). During his years at Iowa, Miles also worked as a pastor at a nearby Friends Church. Miles had been raised as a Quaker and never wavered in his faith. As will be seen in the account of his European trip, the Quaker connection enabled Miles to travel to several locations in Germany that he otherwise would not have been able to visit (at the time of the Miles trip, the United States had not signed a peace treaty with Germany).

Miles's first academic appointment was a temporary one, a one-year replacement for Raymond Dodge, another noted experimental psychologist who became a life-long friend and mentor. Dodge taught at Wesleyan College in Connecticut, but was about to spend a year on leave at the Carnegie Nutrition Laboratory. On Dodge's recommendation, Miles replaced him at Wesleyan. When Dodge's year was up, he urged officials at the Nutrition Lab to have Miles replace him in Boston, and this led to an eight-year stint (1914–1922) for Miles as a research scientist at the Carnegie Lab. There he developed a close relationship with the director, Francis Benedict (1870–1957), and became involved with projects on the adverse effects of (a) alcohol and (b) reduced and inadequate nutrition.

Near the end of his time at Carnegie, Miles began to miss the academic environment he had briefly sampled at Wesleyan. When an opportunity to teach at Stanford University arose, Miles jumped at it and spent the next decade (1922–1932) at the Palo Alto campus. It was in California in 1925 that his first wife died, and where he subsequently met his second wife, Catharine Cox. Cox was a student of Lewis Terman, involved with Terman's famous longitudinal study of genius (Terman, 1925). It was Cox who completed the well-known study in which IQ estimates were made of important historical figures (Terman, Cox, et al., 1930); it was her doctoral dissertation.

Walter Miles flourished at Stanford, developing a research program that, typically for him, encompassed a variety of topics. These included the measurement of eye movements, the study of maze learning in rats (including the invention of several types of mazes), and the assessment of age-related changes in various cognitive and behavioral factors (the Stanford Later Maturity Project). His continuing fascination with research apparatus is evident in his observation that a "piece of apparatus designed to provide a task for a human subject and to give a score or measurable record of his performance seems to me to offer a standing invitation to research curiosity" (Miles, 1967, p. 235). Miles was developing a national reputation among his peers, as his 1932 election to the presidency of the American Psychological Association demonstrates. As a member of the prestigious Society of Experimental Psychologists, Miles won the coveted Warren Medal in 1949 in recognition of the excellence of his research.

In the early 1930s, Raymond Dodge once again entered Miles's life, convincing his colleague and protégée to return to the east coast. Dodge, then on the faculty at Yale University's Institute of Human Relations, was nearing the end of his career and was increasingly disabled by Parkinson's disease. Wishing to insure that his laboratory would be in good hands, he prevailed on the Yale administration to recruit Miles. Catharine was also given a position in the medical school at Yale. The Mileses stayed at Yale for just over twenty years, from 1932 to 1953. Miles retired at Yale's mandatory retirement age of 68, but he remained active—in 1954, just entering his seventies, he took a three-year visiting professor position at the University of Istanbul, successfully establishing a psychology laboratory there (Miller, 1980). Miles died in 1978 at the age of 93.

Miles at the Carnegie Nutrition Laboratory

Steel magnate and philanthropist Andrew Carnegie founded the Carnegie Institute in Washington (www.ciw.edu) in 1902, its purpose being to provide financial support for scientific research. One of earliest scientists to be funded by the Carnegie was the chemist and physiologist Wilbur Atwater (1844–1907) of Wesleyan University in Connecticut. Starting in 1903, Atwater received regular grants from the Carnegie Institute, enabling him to carry on detailed studies of metabolic processes in humans. Much of the funding went into the construction of a highly sophisticated "respiration calorimeter," a chamber large enough to hold a reclining human subject, and the primary apparatus for Atwater's research on metabolism. The device was designed to measure the relationship between energy intake (through food and drink) and the resulting metabolic effects in the body. Atwater was aided in his research at Wesleyan by Francis Benedict (1870–1957), who also had training in chemistry and physiology.

In 1906, the Carnegie Institute significantly increased its support of Atwater and Benedict, approving the building of a separate laboratory for their research—the Carnegie Nutrition Laboratory. It was to be located in Boston, within a block of Harvard

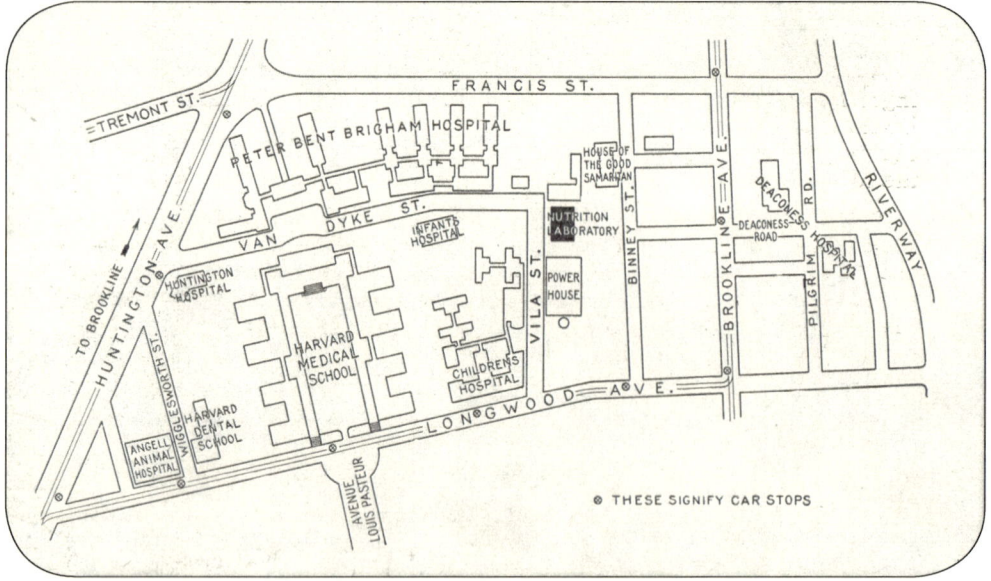

Figure 1. Location of the Carnegie Nutrition Laboratory in Boston.

Medical School and Peter Bent Brigham Hospital. Ground was broken for the construction of a modern three story building in summer 1907 and the lab opened in February of 1908. Figure 1, from a card in the Miles papers, shows its location and provides directions from various points in Boston.

Figure 2. The pursuit meter, created by Miles to measure hand-eye coordination.

Unfortunately, Atwater did not live to see the completion of the Boston lab. He died in 1907, and Benedict was named the laboratory's director. For Benedict, the work of the Nutrition Lab became his life's passion—he directed the laboratory for 30 years until his 1937 retirement. From the start, Benedict developed close ties with physiologists in Europe, making his first tour of European laboratories in 1907. These tours continued with some regularity over the years—two boxes in Benedict papers at the Countway Library of Medicine (Harvard Medical School) contain "Reports of Visits to Foreign Laboratories, seven volumes, 1907–1933" (Finding Aid, Benedict papers). The Miles European visit in 1920, then, was part of the Nutrition Laboratory's general practice of maintaining direct contact with European science.

By 1915, the second year in residence for Miles, the Nutrition Laboratory had a staff of 25 (Nutrition Laboratory, 1915), and it was fully immersed in its program of research to examine "the physiological effects of various nutrients upon the human body" (Benedict, 1915, p. 75). The laboratory housed four different calorimeters, including two large enough to hold exercise equipment (e.g., the 1915 version of a stationary bicycle, called a bicycle ergometer), enabling research on the effects of muscular exercise on respiration and metabolism. The laboratory also investigated "the influence of alcohol upon the metabolic, neural and muscular processes" (p. 82). Miles was deeply involved in this research throughout his tenure at Carnegie, producing several papers (Miles 1916, 1918) and a long monograph (Miles, 1924) on the effects of low doses of alcohol on various physiological and psychological measures. Miles also spent considerable time investigating the effects of "under-nutrition" on human physiology and behavior, showing that diets severe enough to produce a 12% weight loss resulted in a variety of adverse cognitive and behavioral consequences (Miles, 1918).

Consistent with his life-long fascination with apparatus, Miles developed several pieces of equipment in conjunction with his alcohol and nutrition research. These included the pursuit pendulum (Miles, 1920a), pursuit-meter (Miles, 1921), and the ataxiometer (Miles, 1922). The first two required close attention and hand-eye coordination and the third measured body-sway and general steadiness while standing. With the pursuit-meter (Figure 2), the subject had to move a dial to keep a visual stimulus within a crosshair that was constantly moving. It was a forerunner of the modern pursuit rotor apparatus.

During his time in Boston, Miles developed a close and congenial working relationship with Benedict. The friendship lasted well beyond Miles's departure for Stanford in 1922. The strength of the relationship can be discerned from a letter Benedict wrote at the time of his 1937 retirement. In part, it read,

As I look back over the three decades that the Nutrition Laboratory has been running, naturally I think a great deal of our experiences together. . . . When you left to go to California I was literally heart-broken . . .
Quite aside from your understanding, keenness of mind, indefatigable energy, and enthusiasm I have felt that we were in as nearly perfect rapport as any two men could be. . . . The imprint you left upon the Nutrition Laboratory, on scientific activity and on me and my scientific thought in designing powers and even expression has been greater than with any other one man I know. (Benedict, 1937)

When it came time for one of the Carnegie researchers to visit Europe in 1920, it is no surprise that Benedict chose his trusted colleague Walter Miles. The trip was especially important because World War I had seriously disrupted both European science and the normal lines of scientific communication among those in different countries. For Miles the trip was of great value because he would be meeting many prominent scientists for the first time. Also, it would give him a direct look at laboratories in other countries and enable him to see, first-hand and in actual operation, research apparatus he only had read about in journals. As a "lab man," this trip was a dream come true for Miles.

Miles and His Grand Tour of Europe

When Miles sailed for Europe on April 11, 1920, World War I had been over for only a year and a half, and the wounds were still raw. Although the battlefields of the war in Western Europe occupied a relatively small geographical area—large portions of Belgium and northern France—the war produced unprecedented loss of life, devastated the economies of the countries in the region, and the 1919 Paris peace conference perpetuated resentments that led to World War II just twenty years later (MacMillan, 2001). Among the scientists that Miles met on his trip, some lingering hostility was evident. As you will read in his account, for example, it is clear that French and German scientists were not likely to be research collaborators for some time.

The war also had direct effects on the research conducted by the scientists Miles visited. In England, for instance, several of the scientists he met were frustrated by the difficulty they were having in re-establishing their programs of research. Funding was part of the problem, but a major issue was the return of large numbers of soldiers who wished to pursue higher education. In many universities, professors had to cope with large classes and in some cases, the conversion of their precious laboratory space to instructional space. The war also affected the nature of some of the research Miles learned about—several of the researchers had been diverted from their normal research programs to become involved in war-related projects, such as the selection of pilots for the just-developing air forces, the use of sound localization methods to detect artillery positions at the front, or the application of perceptual research to the design of camouflage.

In addition to the Miles *Report of a Visit,* the Miles papers at Akron also include considerable other information about his trip—diaries, calendars, and a long series of letters written by Miles after the trip, following up with all of his new international colleagues. An overview of his itinerary can be discerned from this letter that Miles wrote to Dodge upon his return to Boston. The tour, Miles wrote, included

one month in England, three weeks in Paris, a week in Belgium, two in Holland, a week each in Denmark and Sweden, and shorter times in Switzerland, Germany, and Austria. In the latter two places, I visited practically only the relief work of Quakers ... It was my observation that the Germans and Austrians bore no hatred toward the Americans. (Miles, 1920)

It appears that Miles had some control over his itinerary, being sure to visit as many psychology laboratories as he could. Pre-World War I Carnegie trips to Europe only included stops at laboratories devoted to physiology, pharmacology, and chemistry, or connected with medical schools (even visits to Pavlov's laboratory at the Institute for Experimental Medicine in St. Petersburg in 1907, 1910, and 1913 were considered visits to physiology labs, and Pavlov never considered himself a psychologist). Miles, however,

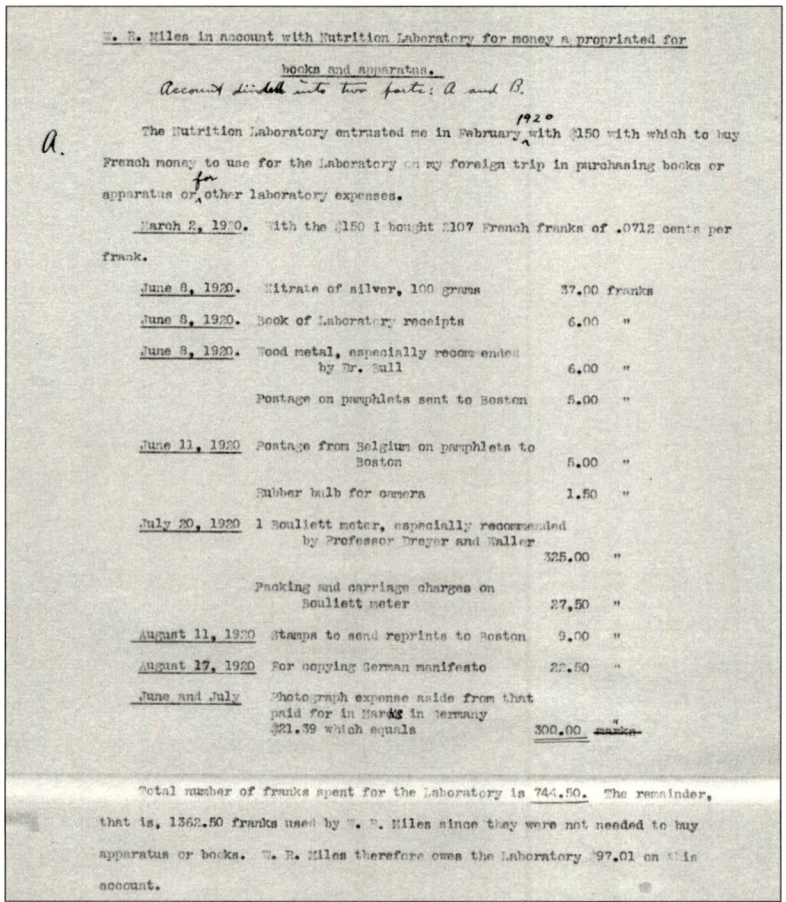

Figure 3. Budgets. Converting dollars to francs and marks.

while scheduling return trips to many of the labs seen in the earlier Carnegie visits, also included laboratories of experimental psychology on his trip—he visited psychologists in London, Edinburgh, Glasgow, Cambridge, Oxford, Paris, Groningen, Copenhagen, and Leipzig. He also attended a meeting of the British Psychological Association.

Miles was known for making meticulous records of all of his activities, and seldom discarding any of his documents (Goodwin, 2003)—the Miles papers at Akron include 128 boxes (70 linear feet) and a highly detailed 756-page finding aid. An example of his attention to detail (and his essential frugality) can be seen in the records Miles kept of his costs during the trip. The Laboratory advanced him $350 to be converted into foreign currency and used to purchase books and laboratory materials during his time in Europe. For example, the documents in Figure 3 shows that Miles used $150 of

INTRODUCTION

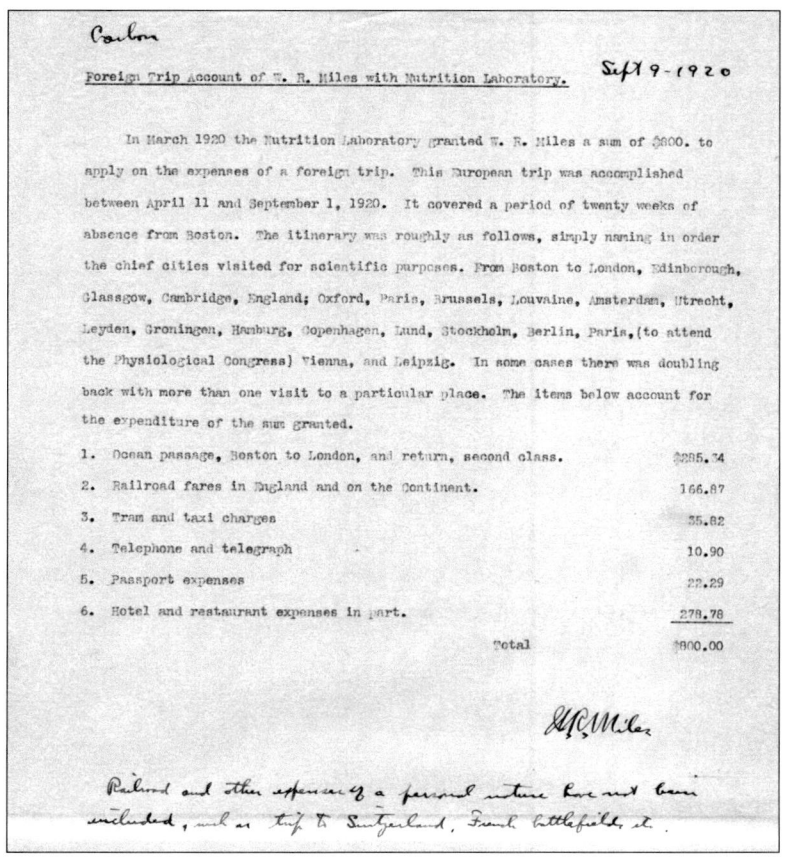

Figure 4. Budgets. Overall cost of the Miles visit ($800).

this money to buy 2,107 French francs and the remaining $200 to buy 18,348 German marks. These funds he spent on items ranging from 100 grams of nitrate of silver to a "Bouliett meter" to photographic film and developing. On his return, Miles paid back the funds he had not spent ($109.44 of the original $350). As for travel costs, Figure 4 shows that the European tour cost the Nutrition Laboratory $800 (just over $8500 in 2008 dollars). That Miles tried to keep his travel costs at a minimum is evident from his decision to sail in "second class" at a cost of about $100 each way.

Costs not related to the trip's primary purpose Miles bore himself—the handwritten note at the bottom of Figure 4 shows that Miles used his own money for side trips. These included a brief and pleasurable visit to Switzerland and a more sobering one to the battlefield at Verdun, France, which he described in his diary: "country swept clean, dreadfully pockmarked ... we go into a dugout and wander their passages ... I find machine gun nest and several rifle shells; hand grenades, rifle barrel,… shrapnel and grenades, gas mask" (Miles, 1920b). Item 6 in the Miles expense page makes it clear that, in order to keep within his $800 budget, Miles used his own money to cover some of the hotel and restaurant costs. He was able to save some money by staying at the homes of several of the researchers he encountered during the trip.

Upon his return to Boston, Miles prepared the following account, and corresponded with many of the scientists he had met during the trip. In many of these letters, Miles included copies of photos that he had taken—his fascination with photography just one more reflection of his love of apparatus and the technology of his day. Many of these photos are reproduced in the report.

One final point about the report is that it evidently was not one of those accounts subsequently filed away, never to be seen again. Instead, handwritten notes in the margins (not in Miles's handwriting) make it clear that the report was readily available to Carnegie researchers and referred to in subsequent years, with information occasionally updated. At several places in the manuscript, for instance, notes appear, usually referring to the subsequent deaths of researchers visited by Miles (e.g., page 248 refers to the 1949 death of August Krogh).

In Sum

The European trip made by Walter Miles in the spring and summer of 1920 was of great importance to the Carnegie Nutrition Laboratory. World War I had seriously disrupted Carnegie's contact with its European counterparts, and it was up to Miles to determine the current status of these laboratories, gauge the effects of the war on research productivity, and reestablish contact, with the eventual aim of rebuilding the kind of collaborative research that had existed prior to the war. That Miles was thorough in his assessment of the laboratories he visited is an understatement, as you will see for yourself as you read through the manuscript. You might consider starting the Miles *Report of a Visit* with the final section. In his "General Remarks" (page 299), Miles briefly describes prior visits to European laboratories by Benedict and other Carnegie staff, explains the reasons why his visit was important, and describes some of the highlights of the trip.

References

Benedict, F. G. (1915). Investigations at the nutrition laboratory of the Carnegie Institute of Washington, Boston, Massachusetts. *Science, N.S., 42*(1072), 74–84.

Benedict, F. G. (1937, August 16). [Letter to Walter R. Miles.] Walter R. Miles and Catharine Cox Miles papers (Box M1124, Folder 1). Archives of the History of American Psychology, University of Akron, Akron, OH.

Finding aid. (n.d.). Benedict, Francis Gano, 1870–1957. Papers, 1870s–1957: A finding aid. Center for the History of Medicine, Francis A. Countway Library of Medicine, Harvard Medical School, Boston, MA. Retrieved from http://oasis.lib.harvard.edu

Goodwin, C. J. (2003). An insider's look at experimental psychology in America: The diaries of Walter Miles. In D. B. Baker (Ed.), *Thick description and fine texture: Archival research in the history of psychology* (pp. 57–75). Akron, OH: University of Akron Press.

Helson, H. (1953, March 3). [Letter to Neal Miller.] Walter R. Miles and Catharine Cox Miles papers (Box 1199.3, Folders 2 & 3). Archives of the History of American Psychology, University of Akron, Akron, OH.

Hilgard, E. R. (1980). Walter Richard Miles: 1885–1978. *American Journal of Psychology, 93*, 565–568.

MacMillan, M. (2001). *Paris 1919: Six months that changed the world.* New York: Random House.

Miles, W. R. (1916). Some psycho-physiological processes as affected by alcohol. *Proceedings of the National Academy of Sciences, 2*, 703–709.

Miles, W. R. (1918). The effect of a prolonged reduced diet on twenty-five college men. *Proceedings of the National Academy of Sciences, 4*, 152–156.

Miles, W. R. (1920a). A pursuit pendulum. *Psychological Review, 27*, 361–376.

Miles, W. R. (1920b). [Diary entry.] Walter R. Miles and Catharine Cox Miles papers (Box 1164, Folder "1920 Tour"). Archives of the History of American Psychology, University of Akron, Akron, OH.

Miles, W. R. (1921). A pursuit-meter. *Journal of Experimental Psychology, 4*, 77–105.

Miles, W. R. (1922). Static equilibrium as a useful test of motor control. *Journal of Industrial Hygiene, 3*, 316–331.

Miles, W. R. (1924). *Alcohol and human efficiency.* Monograph 333. Washington, D. C.: Carnegie Institution.

Miles, W. R. (1967). Walter R. Miles. In E. G. Boring & G. Lindzey (Eds.), *A history of psychology in autobiography* (Vol. 5, pp. 221–252). New York: Appleton-Century-Crofts.

Miller, N. E. (1980). Walter R. Miles (1885–1978). *American Psychologist, 35*, 595–596.

Nutrition Laboratory (1915, February). In *Annual Report, Carnegie Institute of Washington,* Volume 4 (pp. 22–25). Washington, D. C.: Carnegie Institution.

Terman, L. M. (1925). *Genetic studies of genius. Vol I. Mental and physical traits of a thousand gifted children.* Palo Alto, CA: Stanford University Press.

Terman, L. M., Cox, C. M., et al. (1930). *Genetic studies of genius. Vol II. The early mental traits of 300 geniuses.* Palo Alto, CA: Stanford University Press.

Editorial Note

Walter Miles's *Report of a Visit to Foreign Laboratories: April to August, 1920* is part of the Walter R. and Catharine Cox Miles papers housed in the Archives of the History of American Psychology at The University of Akron. The manuscript is reproduced here as an archival facsimile. The original is available for viewing at the archives on an appointment basis.

The manuscript has been reproduced in its original typescript. Miles took numerous photographs which appear throughout the report. He also included postcards and other images, some of which we were unable to reproduce due to copyright restrictions; in some cases, entire pages of copyrighted material have been omitted from this book. When these copyrighted images appear on pages reproduced in this book, their omission has been noted. Transcriptions for Miles's handwritten notes and photograph captions appear at the back of the book. These pages have been modified in the following manner: the scale has been reduced to fit within the pages of this book, and section pages, headings, and page numbers have been added to enable the reader to navigate this work.

GREAT BRITAIN

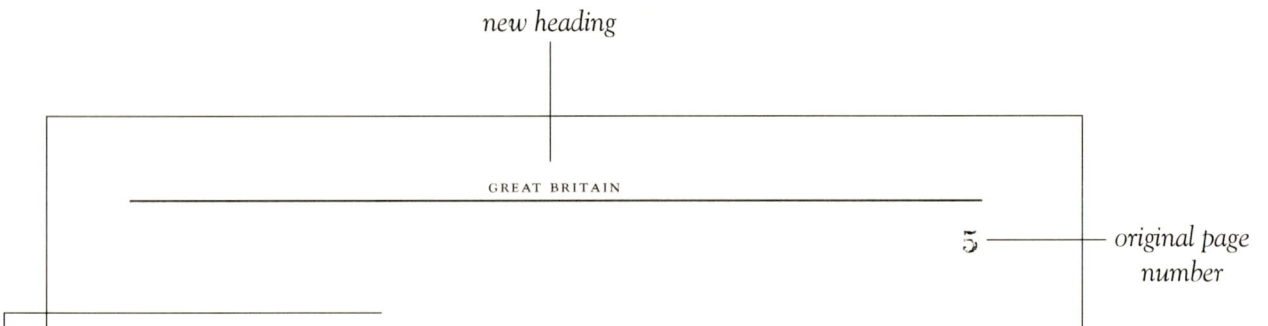

University College, Physiological Laboratory.
Professor Bayliss and Dr. Drummond.

The department of physiology at University College occupies what is known as the Institute of Physiology on Gower Street built in 1909 (See fig. 1). It was planned by Starling and Bayliss, is well adapted to experimental purposes, and is the most complete institute of its kind in London. The department of Physiology includes the subjects of Histology and Physiological Chemistry. The latter subject is in charge of Drummond who has lately taken the place of Plimmer who in turn has gone to Edinburgh. Starling was in India in connection with the establishment of a research laboratory for the British Government. Bayliss commented upon the heavy amount of teaching now necessary in all the British Universities. The war held large numbers of men away from their medical courses, now these have returned and cannot be put off. They will clog the schools badly for the next two or three years. Bayliss teaches histology and his research interests are particularly histological problems. His time recently has been spent in getting out a second edition of his splendid book "Principles of General Physiology". His recent work and writing has been along lines prompted by war problems e.g., a small book on "The Physiology of Nutrition", articles on "An Accessory of Food", "The Perception of Sound", "A Ministry of Health", "Intravenous Injection of Gum in Low Blood Pressure", and "The Psychology of Food and Economy in Diet", etc. As I recall, his most recent article which is more typical of his usual lines of research is "The Mechanism of Chemical Change in the Living Organism". Bayliss, dean of the Faculty of Science assumes also a good deal of the administrative responsibility for the Institute due to the poor health and outside interests of Starling. He has no special office, only a desk-table in one of the laboratory rooms. The picture of him (See fig. 2) shows him sitting at this desk. His quiet effeciency was indicated to me by the fact that he wrote a letter

Walter Miles and His 1920 Grand Tour of European Physiology and Psychology Laboratories

REPORT OF A VISIT TO FOREIGN LABORATORIES

APRIL TO AUGUST, 1920.

By

Walter R. Miles.

Nutrition Laboratory of the Carnegie Institution of Washington.

Boston, Massachusetts,

1920.

This is the carbon copy of the Foreign Report written for Nutrition Laboratory. It contains nearly all the pictures that are in that report and these have been numbered to correspond with the typewritten material. There are many other pictures not directly mentioned in the typed copy.

4 visits on Grand Eur. Tour

London - Spearman p.13.
 Lecturing + research
Wants H. Wynn Jones lectures on senses &
mental tests   nervous system
Brit. Psy Assn - Scripture 57
   Paper read on Speech inscriptions in
                        normal & abnormal conditions
Edinburgh Drever   good lab. 67
  Working on problems of social ψ
  at variance with McDougall
Had been **Glasgow** Watt interested 73
student   in your search
of Wundt. Felt attitude of Prof. very arbitrary
Later worked with Külpe a diff. type encouraged freedom
                                                   of view
Cambridge Bartlett 87
Bicycle Ergometer Adrian = Lucas work
Oxford   McDougall 110
Favored use of moderate amt. alcohol. This
might improve effect at only small loss in
                                        effect.
France
Paris Piéron 127

Netherlands
Groningen Heymans 205

Denmark Lehmann 213
Copenhagen (absent) doesn't speak
   Lab ok on Carpenter's visit English.
Germany ʒ Kruger   242
Leipzig   ʒ Kirschman

ITINERARY AND INDEX.

City	Institutions	Investigators	Page
London	University College (Physiological Laboratory)	Bayliss	5
		Drummond	9
	(Psychological Laboratory)	Spearman	13
	University College Hospital	Lewis	19
		Cotton	19
	King's College Medical School	Halliburton	24
	Guy's Hospital Medical School	Pembrey	28
		Ryffel	33
	Imperial Institute (Physiological Laboratory)	Waller	34
	University of London Club: Lecture	Bose	43
	King's College for Women	Mallenby	45
	Ministry of Health, Whitehall, London	Newman	47
	National Institute for Medical Research, Mount Vernon	L. Hill	48
		Moore	51
		Dale	52
		Brownlee	53
		Schuster	53
	Meeting British Psychological Association, Bedford College, Regents Park	Scripture	57
		Klein	58
Edinburgh	University of Edinburgh (Physiological Laboratory)	Schafer	59
		Taylor	61
	(Royal Infirmary)	Meakins	64
	(Pharmacological Laboratory)	Cushny	66
	(Psychological Laboratory)	Drever	69
Glasgow	University of Glasgow (Physiological Laboratory)	Paton	71
		Cathcart	71
	(Psychological Laboratory)	Watt	75

City	Institutions	Investigators	Page
Cambridge	University of Cambridge	Barcroft	77
	(Physiological Laboratory)	Adrian	80
		Hartree	85
	(Psychological Laboratory)	Bartlett	87
		Muscio	88
	(Bio-chemical Laboratory)	Hopkins	90
		Peters	90
		Grey	90
	Meeting British Physiological Society		92
	Cambridge and Paul Instrument Co.	Whipple	94
Oxford	University of Oxford	Sherrington	96
	(Physiological Laboratory)	Bazett	100
		Douglass	102
	(Haldane's Private Laboratory)	Haldane	106
	(Pathological Laboratory)	Dreyer	109
	(Psychological Department)	McDougall	116
Paris	Marey Institute	Bull	117
		Nogues	127
	Physiological Laboratory of the Sorbonne	Lapique	129
	Psychological Laboratory of the Sorbonne	Pieron	134
	Pasteur Institute	Bertrand	137
	Societe Scientifique d'Hygiene	Alquier	139
	Alimentaire	Lefevre	139
	Physiological Congress of 1920 (Germans and Austrians excluded)		144
Brussels	Solvay Institute	Heger	150
	(Physiological Laboratory)	Phillipson	150
Louvain	Catholic University		
	(Physiological Laboratory)	Noyons	152
Amsterdam	National Institute for Nutrition	Van Leersum	167
	University of Amsterdam (Neurological Institute and Laboratory)	Salomonson	171
	Free University (Department of Biology)	Buytendyk	177
		Dirken	177

City	Institutions	Investigators	Page
Utrecht	State University of Utrecht (Institute of Physiology and Physiological Chemistry)	Pekelharing Zwaardemaker Ringer	183 183 186
	(Institute of Inorganic Chemistry)	Cohen	187
Leyden	State University of Leyden (Physiological Laboratory)	Einthoven	188
	(Pharmaco-Theraputic Institute)	Van Leeuwen	193
	(Physical Laboratory)	Onnes Crommelin	196 196
Groningen	State University of Groningen (Physiological Laboratory)	Hamburger	198
	(Psychological Laboratory)	Heymans Brugmans	205 205
Copenhagen	University of Copenhagen (Animal Physiological Laboratory)	Krogh	208
	(Physiological Laboratory)	Henriques	216
	(Psychophysical Laboratory)	Lehmann	217
	Veterinary High School	Mollgaard Anderson	218 218
	Dansk Maalerfabrik	Gjellerup	221
Lund	Karolinska University (Bio-chemistry Laboratory)	Widmark	222
	(Physiological Institute)	Thunberg Westerlund	226 226
Stockholm	Karolinska Institute (Physiological Laboratory)	Johansson	228
	Stockholm Board of Health and Technical High School	Sonden	233
	Military Hospital	Stenström	234
Hamburg	University of Hamburg (Physiological Laboratory)	Kestner	236
Berlin	University of Berlin (Physiological Laboratory)	Rubner Gildemeister	238 239
	(Institute for "Arbeitsphysiologie"	Thomas	240

City	Institutions	Investigators	Page
Leipzig	University of Leipzig (Institute for Experimental Psychology)	Kruger	242
		Kirschmann	242
	(Physiological Institute)	Garten	244
Vienna	University of Vienna (University Hospital "Kinderklinik")	Von Pirquet	247
		Schick	247
	General Remarks		253

Great Britain

University College, Physiological Laboratory.
Professor Bayliss and Dr. Drummond.

The department of physiology at University College occupies what is known as the Institute of Physiology on Gower Street built in 1909 (See fig. 1). It was planned by Starling and Bayliss, is well adapted to experimental purposes, and is the most complete institute of its kind in London. The department of Physiology includes the subjects of Histology and Physiological Chemistry. The latter subject is in charge of Drummond who has lately taken the place of Plimmer who in turn has gone to Edinburgh. Starling was in India in connection with the establishment of a research laboratory for the British Government. Bayliss commented upon the heavy amount of teaching now necessary in all the British Universities. The war held large numbers of men away from their medical courses, now these have returned and cannot be put off. They will clog the schools badly for the next two or three years. Bayliss teaches histology and his research interests are particularly histological problems. His time recently has been spent in getting out a second edition of his splendid book "Principles of General Physiology". His recent work and writing has been along lines prompted by war problems e.g., a small book on "The Physiology of Nutrition", articles on "An Accessory of Food", "The Perception of Sound", "A Ministry of Health", "Intravenous Injection of Gum in Low Blood Pressure", and "The Psychology of Food and Economy in Diet", etc. As I recall, his most recent article which is more typical of his usual lines of research is "The Mechanism of Chemical Change in the Living Organism". Bayliss, dean of the Faculty of Science assumes also a good deal of the administrative responsibility for the Institute due to the poor health and outside interests of Starling. He has no special office, only a desk-table in one of the laboratory rooms. The picture of him (See fig. 2) shows him sitting at this desk. His quiet efficiency was indicated to me by the fact that he wrote a letter

Fig. 1. Institute of Physiology, located at Gower Street. This building faces on a court. The physiology work was mostly in the right hand end. The library and "Tea Room" were above the entrance. In the Tea Room a great collection of the photographs of physiologists trained here. Under the entrance is a tunnel way with doors opening into basement rooms and leading to the back where the animal cages are located. Physiol. Chem. is in the left hand end of the building.

Fig. 2. Professor W. M. Bayliss at his laboratory desk in the Physiological Institute, University College, London.

Fig. 4 Group of research students, Lovegore, Anrep and Daly, in Professor Starling's Laboratory.
Note the overhead supply of power, gas and air. Anrep was trained with Pavlov.

Dr. Gleb V. Anrep

Fig. 3. Dr. I. de Burgh Daly experimenting with the string galvanometer and audion valves.
This was an Edelmann galvanometer with a Cambridge string carrier, a combination which seemingly worked well.

while I arranged the camera and focused it.

One of the most interesting rooms of the Institute is the "tea room" on the walls of which are many pictures of historical interest for physiology and also pictures of all those who have taken degrees in this department, it makes up a surprisingly large and important group.

During my visit with Bayliss Sir George Thane, general supervisor under the Home Office for the Anti-vivisection Act in England made a call. Sir George was previously a Professor at University College. (A picture of him with Dr. Pembrey is fig. 8.) In connection with Sir George's visit I learned considerable about the Anti-vivisection Act requirements. In all instances it is necessary to get a license to use animals for vivisection purposes. Application for this license must be signed by two responsible people in the department or faculty. It requires about three weeks to secure it and the number of animals is limited especially if the operation is to be made and the animal kept alive subsequently. It is particularly difficult to get a permit to use dogs and especially in such an arrangement as employed by Pavlov of Petrograd. Bayliss was speaking with Sir George that they should be allowed a permit for Dr. Anrep, a Russian, who has previously been an assistant of Pavlov. Bayliss thought the appeal could be made on the basis that Anrep already knows the technique having worked with Pavlov. The Anti-vivisection group in England are continually active, being led, as I gather, by a Dr. Collins and Dr. Headwin, both of whom Bayliss termed "the universal antis". They are very eloquent and strong backers of complete anti-vivisection, anti-vaccination, etc., etc.

Another interesting feature of my visit was in reference to the Hindu scientist, Professor J.C.Bose, lately knighted. When Bayliss and I were in the basement in one of the dark rooms, we found several potted plants on a table. He explained their presence by saying that it was in that room that Bose had lately demonstrated his cyscograph to a committee composed of Dr. Bayliss, Sir Oliver Lodge, and several other very distinguished people. He told me something about the nature of the demonstration made on plants. (I shall have more to say on a later page concerning Bose and his work on plant physiology as it has at present interested English scientists.).

Drummond had class work in progress at the time of my visit and my conversation with him was very short. I simply saw the laboratory in which physiological chemistry was being taught.

Anrep, formerly a student and assistant of Pavlov, was working in the physiological laboratory. Regarding Pavlov, he made the following remarks. "I saw him last in March, 1916. At that time he had a broken femur but recovered the accident. He was in a bad condition physically. His work also was practically at a standstill. He has a book on "The Conditioned Reflex" all written but it cannot be published. We all went to the laboratory in those days but could do no work. All the dogs died for lack of food. At one time Professor Pavlov had 25 dogs each giving one quart of gastric juice per day. A dog could give two quarts but that was too much for the dog to continue a long time. Giving one quart a day he could live indefinately. Metabolism measurements were never made on these dogs who were used for the experiment of mock feeding and collection of gastric juice. (In speaking of Pavlov's dogs, Anrep said that Pavlov gave the dogs some acid. The loss of so much gastric juice greatly increased the alkaline reserve. He remarked that the bones of the dog would in time get soft and the dog would be badly bow-legged.) Finally only five or six dogs were living. Pavlov quit lecturing and his professorship in the Military Medical Academy and did all his work in experimental medicine and entirely in the laboratory for Conditioned Reflex. News through Arrhenius of Stockholm rather recently stimulated the British Royal Society and others in London to gather up money in December of 1919 to be sent to the Danish Red Cross for the relief of Pavlov." (I later learned from conversation with Arrhenius in Stockholm that this money, 2000 crowns or more, had not been delivered and he had nothing further in reference to Pavlov.) Anrep was working at present on problems of digestion but his research was distinctly hampered according to him by the Anti-vivisection Act

He had recently finished a problem on the pitch discrimination of dogs finding that with tones ranging from 635 to 652 V D the dog could discriminate a difference of 1/8 tone. This is not in agreement with the work of Johnson formerly of Nela Park, Cleveland. Anrep was just finishing a paper on "The Metabolism of the Salavary Gland". He is in a rather difficult situation, having been a recent assistant of Pavlov, he naturally knows about the recent researches in those laboratories. He himself is keen to continue work in this field but does not feel at liberty to publish because much of what he would say is the result of the work in Pavlov's laboratory and Pavlov has had no opportunity to publish or in any way describe these researches up to the present.

Bayliss told me that at present he was trying to do some work on the toxiticity of hemoglobin. He thinks that pure hemoglobin is likely not toxic. He is also working upon the permeability of the bladder for certain electrolytes. Then there is the problem that he is undertaking for the government in reference to freezing of foods, especially beef, which must be only chilled, and at present can be sent from Argentine but not from Australia to England. They had a special installation in the basement to do this work. He was interested in my telling him something about the Jewish customs in reference to beef in the United States.

Daly is doing research on electrocardiogram outfits. At the time he was especially trying to employ audion amplifyers, i.e., Fleming and Marconi valves, to amplify the deflections of the string galvanometers His desire was to simplify the apparatus for chemical work so that a less sensitive instrument would be satisfactory and less trouble experienced with breaking and mounting strings in the galvanometer. Something of his arrangement is shown in figure 3. He had a combination of Edelmann field coils and pole pieces fitted with a Cambridge string holder and was using the Cambridge falling plate camera. I informed him of the work being done at the Harvard Medical School by Forbes along the same lines using the audion valves to amplify deflections from action currents of nerves and other weak sources of E. M. F.

Daly during the war was a flying pilot for two years then an instructor and examiner of prospective pilots for two years. It was his opinion that for the man who has not learned to fly, quickness in reaction time is a good test of his probable ability, but for the fighting pilot reaction time is not a satisfactory measure for his proven ability. "You must know from the medical officer" he said "what he is like, whether he is drinking much alcohol, smoking a great deal, how he plays billiards, how much he raises you in a poker game, etc. You must take account of all such things to keep track of his condition." Daly was interested in the pictures and diagrams of the pursuit pendulum and the pursuit meter. He raised the question, what could a practiced subject do on the pursuit meter when he was normal, when he was hypnotized, and when under post-hypnotic suggestion. He was much interested in this field of research and thought the pursuit meter quite adaptable as an objective means of studying these different conditions.

University College, Psychological Laboratory,
Professor Spearman.

Spearman's laboratory is open from 3 to 7 P. M. in the afternoon. All of the teaching, and I understood, most of the research work is done between these hours. At other times the laboratory cannot be seen. I was fortunate in arriving there just in time for tea, in fact, Spearman had invited me to tea over the telephone. We had pleasant conversation about psychology and psychologists. They were interested to see the pictures of some psychologist taken at Cambridge, Mass., December, 1919, which I happened to have with me. Associated with Spearman is Aveling, who holds the degree of Ph. D., D. Sc., and D. D.? from Louvain, London, and Rome. His interest is in the history of psychology and on the philosophical side. Jones (Ll. Wynn) from North Wales, is another associate and is lecturing on the senses and the nervous system with demonstrations. He is especially interested in securing mental test materials also some apparatus as he is starting a psychological laboratory in the University of Leeds next October, 1920. He is a very charming fellow and it was he who showed me all about the laboratory it being the first day of the term and Spearman was very busy in conference with students. There were two other men, the latter working for a degree, whom I met and was well impressed with, Dr. Flugel and Captain Philprit.

Their total laboratory space would equal about 5 of the usual laboratory rooms. Their department seemed to me quite cramped for space. There was the Sharpe Lecture Room, a general laboratory room occupied for research work by Spearman, another room divided into two sound-proofed (?) rooms, an office for Aveling, a fourth room divided into two dark rooms used for research purposes and Spearman's office.

Fig. 5. "Tea Time" in the Psychological Laboratory, University College, London. Professor Fr. Aveling, Professor C. S. Spearman, and Capt. Philpot.

Professors Spearman and Aveling. Taken at a guess on the focus, as I found the ground glass screen had been left in the library of the Physiology Institute. Went and recovered it before taking Fig. 5 and never left it again. At Brussels I completely failed one picture of Professor Heger and myself by leaving in the slide of the screen focus camera — to my great regret.

Fig. 6 – Professor W. D. Halliburton in his office at King's College Medical School, London.

During the war and lately Spearman and Jones have been cooperating on a piece of research on the subject of "Stereopsis in Aviation". The problem which the aviation subject or prospective pilot has before him is really one of pursuit but involves accurate accommodation of the eye for different distances. Two small movable wagons on parallel toy railroad tracks are arranged so that one is moved by a motor and two eccentrics. The total course of movement is about four feet. The other moves lightly by a hand belt operated by the subject who sits in a position at the end of the track about ten feet away and looking through a slit with one eye sees two colored plates illuminated from behind and carried respectively one on the wagon moved by the eccentric and the other on the wagon moved by the subject. It is his task during the test to keep these two colored plates or colored fields directly opposite each other, i. e. at equal distance from himself. One wagon carries a platform on which a piece of paper is placed to take the records. The subject's car carries a pencil which writes on the coordinate paper carried by the other car. The platform carrying the corodinate paper is given a slow motion at right angles to the movement of the car. A perfect score would be a straight line across the coordinate paper but they usually get a line which looks something like a pursuit meter record and they count the number of times the line drawn by the subject's pencil crosses the theoretical line on the coordinate paper. They also measure the different distances that the subject's pencil gets away from the theoretical line measuring the extreme height of each loop and multiplying by the number of times such loops occur thus combining the size and number of errors into a total score. They expect to do more work with the apparatus before the research is published. They have conferred with certain of their military authorities with reference to the problem and have their cooperation.

An apparatus to measure gun pointing ability was arranged quite simply. On a drum about 8 inches in diameter a wavy line has been drawn encircling the drum. This line is seen by the subject through a narrow horizontal slit exposing but a short section of it at any moment. This slit is viewed through a telescope with cross hairs and the tip of a pointer easily movable is to be kept in register with the wavy line, and also seen in the telescope. The apparatus makes a graphic record directly compared with the standard line to note correspondence. They have no quantitative measure for accuracy.

An ingenious method for testing the quickness and keenness of observation was arranged by having an iron ball rolled down to an inclined tube, jump out at the end, and fall on to a ruled paper. The subject, who stands at a distance sees the complete paper but not the end of the tube and must tell by watching the ball, (which, of course, moves rapidly,) where the ball strikes on the paper. The correctness of the subject's judgment is controlled by an actual record made on carbon paper under the ruled paper. The platform is graduated and is each time given a slight movement at right angles to the direction of the moving ball. Thus the dots on the carbon paper are not confused with the order of the experiment. The subject calls off the number of the line on which he thinks the ball has landed. This is recorded and later by the carbon paper record the direction and the amount of error may be determined. The test seemed quite a good one. However, it does not proceed rapidly and a man can tell by the distance the ball rolls in the tube about how far it is going to jump when it leaves the tube. This secondary criteria for judgment may probably be nearly offset by starting the ball over the same place in the tube and changing the angle of the tube from the horizontal. The experimenter must do his part of the work behing a screen.

Philprit was working on the problem of choice reaction. He had a variety of stimulae, sound of a bell, flash of light, a tap on the finger, and three reaction keys. The hand of the subject lay at one side preliminary to reaction and upon receiving the stimulus the hand is lifted and the finger is passed down a hole to touch one of the keys. His use of the choice reaction is to examine the work curve. He takes only the reaction errors as his measure. He

did not know of the work of Seashore or of Seashore's instrument, the

Psychoerogograph. Philprit has a good recorder for telegraphic tape. It is made by the Cambridge Instrument Company. He is working also on the circulatory changes in a trephined subject, a man.

Miss Newmark, one of Spearman's students, is doing research on the Psychogalvanic reflex. She has a slow-moving coil galvanometer and the throw from the galvanometer is quite long. The scale was hinged to the wall the same as the one in my laboratory. I had never seen another so arranged. She is following Waller's method and applying the electrodes to the palm and back of the head. Is working on children and wants to correlate the actual amplitude of the psycho-galvanic response as shown by the galvanometer with other measures such as speed and accuracy in choice reactions. With her it is a quantitative rather than qualitative problem. We discussed the difficulties of so applying the electrodes to different subjects at different times as to make the difference in the apparent size of the galvanometer excursion meaningful in terms of emotion. For a mode of stimulas she was going to use an auto horn, the chair on which the subject sits was arranged to drop down from some false legs, she also uses some disagreeable taste-odors but not the electric shock.

I had the pleasure of calling at Spearman's laboratory three times. He had not previously understood the technique for recording the eye-movements by the use of photographic methods. He showed much interest in the photographs of apparatus at the Nutrition Laboratory and we discussed particularly the problems of measuring pursuit movement. Jones was very anxious to learn from me the addresses of individuals and shops from whom he could get apparatus and testing blanks preparatory to establishing the laboratory at Leeds this October.

I was glad to learn through an outside source that Spearman and Professor Karl Pearson were about to cooperate in some investigation. These two very capable men have not been on friendly terms of recent years and it is gratifying that they have at last made up their differences. It makes a more agreeable situation for the students in both departments. I tried to call on Pearson at the Galton Laboratory and was refused admittance by a woman porter. The details are of possible interest and given without any feeling of animosity. I learned that anyone who wished to visit the laboratory should make an appointment in advance. Hence, I telephoned one day to University College and asked to be connected with the Galton Laboratory. They informed me that the Laboratory had no telephone but that they would carry a message for me. I gave my name and said that I was a scientific man from American and desired to visit the Laboratory to see their equipment of computing machines and such accessory apparatus. I told them I should like to come in the afternoon at about 3 o'clock two days hence. They assured me that no other arrangement need be made, that I would be quite welcome as they would certainly deliver the message. I was very punctual in reaching the Eugenics Laboratory on the day and time appointed. The porter was absolutely firm in her position that no one could come in or visit the laboratory unless he had with him a letter from Dr. Pearson showing in writing that he had an appointment. She said that Dr. Pearson was a very busy man and could not be interrupted. I told her fully the circumstances of my telephoning for an appointment that I would have been glad to write and would be glad to write for an appointment at that date were it not for the fact of my leaving for Edinburgh the following day. I said that while I should have been very glad to meet Dr. Pearson, I would be happy if any assistant could take the time to show me some of their computing instruments. This was also refused. The porter would not even take my card with a written statement on it that I had called and was exceedingly sorry not to have made a written appointment and so not to have been able to visit the laboratory. I have no doubt Professor Pearson's rule saved him much time and is partly accountable for the great volume of his publications. However, I do not imagine he expects it to be applied so inflexably and upon every instance. Of course, I made no complaint or said nothing about the experience to others. *See letter from Dr. Pearson, next page*

(Copy)

Department of Applied Statistics,
University College, University of London.
The Francis Galton Eugenics Laboratory.   The Biometric Laboratory.

December 1st, 1920.

Dear Sir,

The rule of this Laboratory is that no visitor be admitted without (1) a written appointment being made with me beforehand, (ii) a very definite statement, showing that the object of the visit will repay the loss of time involved, (iii) a written introduction from someone _personally_ known to me. These rules are stringently necessary, because we should otherwise have to pay the cost of an assistant, who would be solely occupied with such visitors. I have known weeks when visitors have called every day, and taken members of the staff off their proper work, and our funds, which are provided for research work, do not permit of such waste of our time and energy.

Had Dr. Miles come with an introduction from Dr. J. Arthur Harris, and written beforehand to state the purpose of his visit and ask for an appointment, I should have done my best to arrange it for him, but the Laboratory is here as a work-place and not as a show-place, and we must maintain these regulations.

I have absolutely no objection to your publication of Dr. Miles' statement, especially if you add to it my present explanations.

As for the telephone message, we have no telephone in the Laboratory, and we cannot be responsible for messages not brought to us, and not replied to by us.

I am,

Yours faithfully,

(Signed) Karl Pearson.

Prof. Francis G. Benedict

Carnegie Institution of Washington.

University College Hospital, Heart Clinic.

Doctors Lewis and Cotton.

Dr. Lewis was very cordial in his invitation over the telephone that I should visit his clinic. Some time ago he had visited the Nutrition Laboratory in Boston. I had with me a letter from Dr. S. A. Levine of Boston who worked in Lewis' heart clinic several months during the war. The clinic is in a large basement room. Here I found about 30 male patients nearly all of whom were stripped to the waist for chest and back examinations. Of these men 25 were patients returning for observation. These observation periods were a month apart and the re-examination was made by Dr. Cotton. It was largely an examination by stethescope and having the subject breathe according to directions. Each subject was asked if he were gaining or losing weight. Cotton insisted that they must weigh themselves frequently. Most of the patients were cases of "soldier's irritable heart" and were government pensioners.

Lewis was interested to hear from Levine and after reading the letter, we talked about Levine's newly established heart station in Boston. We discussed using the string galvanometer during actual physical exercise. Lewis said he had never done this, but that of course the thing you want during actual exercise is pulse rate and also blood pressure. The latter you can hardly get during the work. He wanted to know the fastest rate I had ever secured during work. It so happened I had never made examination of this particular point to see how fast a heart rate could be produced. He said that one can produce a rate of 180 per minute with 10 minutes work in normal man.

We turned to the examination of the seven new cases sent him that morning. In examining, the case was stripped to the waist. There were no dressing booths. There was one nurse who had charge of case records but mostly was engaged with hose patients who were returning for re-examination. I thought the room seemed quite cold and noticed a great deal of shivering on the part of the patients es-

GREAT BRITAIN

20

Brown, Shipley & Co.
123 Pall Mall, London.

Image Not Reproduced

25

pecially when they were touched by the hands of examiners. This aggravated some nervousness which they felt at being examined. This was my only criticism of conditions. A young physician took the dictation of Lewis during his examination. Several young physicians were present.

Previous to the examination and before Lewis came to the clinic the electrocardiograms of the patients had been taken by an assistant. The subjects were in a sitting posture. They did not even use a steamer chair. The electrodes were not wound about the wrist and ankle but the subject simply placed his foot in a pan in which was a large pad of wet gauze and placed his hands on such gauze in the hand electrodes. The hands or foot were not washed or especially prepared to decrease the tissue resistance. Lewis did not examine the electrocardiograms before his examination of the patient or in connection with this examinations at least, while the patient was present. In other words, he made his entire diagnosis so far as I could see from anything that was said or dictated to his amanuensis without the help of the electrocardiograms. The routine of an examination was in outline as follows; (1) Inspection of the case on his standing after walking across the room when he was called. Is there any strong palpitation about the region of the heart or in the vessels of the neck, and the arteries at the elbows? Is there clubbing of the fingers, blueness of hands? By percussion locate the bounderies of the heart and state it in reference to the position of the ribs. (2) Subject lying down on a hard narrow couch. Listen over heart with stethescope. Are there any murmurs such as the "to and fro"murmur" or fidulation double top pulse, regurgitation, etc., etc. Is spleen or liver enlarged? Condition of skin, any breaking down of the small blood vessels, any bits of red shown in these vessels which would be bits of heart muscle thrown off and stuck in the small blood vessels, any breaking out on the skin. (3) Subject in a sitting position. Examination of the subject's back by the stethescope and with the subject breathing according to direction. (4) Subject standing for exercise test. Makes ten hops on each foot. If this amount of exercise does not markedly increase the palpitation, subject is told to lift two 10 pound dumbbells from the floor to a position over the head, full arm extension. Thirty times is about all a normal man out of training "wants" to do this. The amount of accelerations of the heart beat and the strength of the beat is gauged by the palpitation over the region of the heart and in the neck and is observed clinically, Lewis naturally having standards in his own mind for classification of the patients. (5) Subject questioned about his occupation, hours of work, conditions at home, and I thought advised most skillfully and with considerable sympathy. (6) As stated above, the electrocardiogram was taken but not consulted during diagnosis. (7) An X-ray examination, fluoroscopic observation, is made in certain cases, there was one among the seven on whom it was necessary. Another one of the seven cases was in an extremely bad condition although he was up and about, a single glance at the man would not reveal his being in a critical condition, however, Lewis thought he might pass out at any moment, for together with his heart disease, he had double pneumonia. In answer to a question about the source from which the patients came, their number, whether they usually had as many, Lewis replied, "The patients are mostly government pensioners, are sent to us by government authorities, and it looks as though we would carry on this way for the rest of our days."

The galvanometer room is directly off the clinical room and is both galvanometer and dark room combined. A screen wire of large mesh has been tacked all over the wall for electrostatic screening. Lewis thought it likely did no good. He uses the complete Cambridge Instrument Company's outfit and has the large Siemens and Halske form of electric lamps. He used the falling plate camera, has a very large tuning fork to run his time marker. He said, "Plates are much better for careful measurement than paper and of course they make possible any number of prints." Lewis takes the three leads of the electrocardiograms on one plate. I saw a specimen of pathological human heart floating in the sink together with the plates which were being washed. The electrodes are the Cambridge Instrument Company's type and are out in the clinic room. A large amount of saturated cloth is placed in the vessels, the hands and foot rest on this cloth without any special preparation and with the subject sitting more upright than in a steamer chair. The Cambridge double-string holder is employed, using of course only one string for the electrocardiograms. In experiments he told me that he used one string for the standard work placing it in the centre of the magnetic flux and using the other one for a signaloor for a myogram, i.e. a companion tracing to give points of reference rather than true curves. The double converging prism is used to bring the shadows of the two strings close together on the plate.

Lewis had this day, April 29, 1920, received the first copy of the new edition of his book entitled "The Mechanism and Graphic Registration of the Heart Beat". He showed me records and work on the relation of the demarcation current to the mechanical response. He thinks the action current precedes the mechanical contraction by about 0.02 seconds and said in reply to something I remembered Stenström to have told me about recent work in progress in Einthoven's laboratory, "I sincerely hope Einthoven has not found that mechanical change comes first." Lewis made his research by sewing several bristles into the auricle of the heart and these bristles made shadow records by the side of the string syncronous with ~~with~~ taking the electrocardiogram. This work had been done with the assistance of two Americans, Drs. Feil and Stroud. To the latter I was very much attracted and we saw considerable of each other in London and later in Edinburgh and tried to meet at Einthoven's Laboratory in Leyden but could not connect. Stroud was working in Lewis' laboratory about two months as I recall. He is now a member of the staff at the Pennsylvania Hospital in Philadelphia.

Lewis showed me the Lucas comparator which is made by the Cambridge Instrument Company and is for the purpose of carefully measuring photographic or kymographic curves. In reference to the careful measurement of the galvanometer's deflections shown by the electrocardiograms, Lewis made the following comment. "You must do all such work yourself, no assistant can be trained to do it. They have not the discretion, they do not realize the importance of small errors. Then too you must do it to get thoroughly saturated with your subject; you must have the greatest familiarity with your plates." He fingered quite affectionately a pile of plates on which he had been working. "It is tedious, I read plates until I almost go blind but there is no other way. I find three pulse waves on one record enough to average and I can check my readings exactly." Dr. Lewis has just published an article on the use of the comparator with the electrocardiogram (See the Journal, Heart, Vol. VII, No. 3, 1920, page 117.) The work in cooperation with Feil and Stroud is also published in this number of Heart. Lewis gave me a copy of his paper entitled "On Cardinal Principles in Cardiological Practice" from the British Medical Journal, Nov. 15, 1919. He invited me to stay with him to lunch but I had another engagement.

I was told by Stroud that some Americans, he gave me the names, had made very poor impressions in visiting Lewis since they had visited him apparently chiefly to tell him how they were in the habit of doing certain things in reference to running such clinics. There is of course a difference between Americans and Englishmen in the freedom with which they talk about their work. Lewis represents an unusual combination. He is really just a young man. He has tremendous enthusiam and willingness for hard personal research and coupled with this a great amount of clinical practice. He is therefore in a position to more exactly appreciate the real value of the string galvanometer technique and other techhniques to the actual practice of medicine.

King's College Medical School, Physiological Laboratory,
Professor Halliburton.

Professor Halliburton, like several other prominent physiologists, is dean of a college, and in consequence has many administrative duties. When I called at the appointed hour, he asked to be excused for a moment for a short conference. His office was at the end of a long hallway which seemed a combination of library and laboratory with the rooms for Physiological Chemistry and Histology opening off. His office was lighted by a large skylight, it did not contain many books, and for a desk, a simple table, indicating that probably most of his work is done elsewhere. When Halliburton returned he began conversation by saying, "I am sorry about that review." meaning the review in Physiological Abstracts, Volume   Page . The fact is, I had not thought of this review in coming to visit Halliburton, and although taken by surprise, did my best to silence him on the topic. He showed clearly his sincere regret. He said, "I had you visualized as an old man on whom something had been put over but even so, I should be impartial and serious, and it was a grave mistake. Why does it happen that men speak so freely?" We discussed the topic at some length. He spoke about having just received a bundle of reprints from the Nutrition Laboratory which he would look over for review purposes soon. I complimented him on "Physiological Abstracts", especially on the quantity of material which he himself covered. He assured me that he put his best into it. We looked about the rooms of the department which includes Histology and Physiological Chemistry. He complained a great deal about the over crowding caused by the men who were held out on account of the war and who have now returned in a large bunch. He said that practically nothing could be done except the teaching and it could not be well done for lack of space. He was doing only a slight amount of research continuing his study of vitamines but he gave no details as to the particular phase of that problem which was taking his attention. The building and equipment are quite old. I saw no special research apparatus for metabolism work.

Halliburton very kindly took me to lunch in the faculty lunch room in the basement. As we went toward it down some stone steps, he said, "This is the path we used to take during air raids which sometimes occurred during the day. Chittenden and Lusk were often here. At night they thought best not to stay in the same room in the hotel. He described in some detail the air raid conditions. In the lunch room I was introduced to Sir Bernard Pares, Professor of the Russian language at King's College. He requested Halliburton to supply him with a list of Russian physiologists who should be brought out of Russia or some relief sent to their assistance. Mention was made by Halliburton of Pavlov and the funds raised for him which had not yet been delivered. Sir Bernard thought he might be able to do something in reference to this matter. After lunch Halliburton and I continued our discussion for a short time in his office and after taking a photograph of him at his desk, I excused myself, understanding well the amount of work that was waiting him since it was the first part of the term and there were people waiting for conferences. At other times, both at the Meeting of the Physiological Society in Cambridge and at the Physiological Congress at Paris, Halliburton demonstrated in several ways his interest and regards for me and the workers of the Nutrition Laboratory.

Following my very pleasant conference with Halliburton, I went, at the invitation of Dr. Augustus D. Waller, to a meeting of the Royal Society in Berlington House. Unfortunately for me, something came up which hindered Waller from being present. however, I became acquainted with Barcroft of Cambridge, who, at our meeting rather took me off my feet by saying at the moment of introduction, "I am going to make a rank Americanism, and do it first, I am glad to meet you." Barcroft suggested that I come to Cambridge to visit him on May 3rd and kindly insisted that he meet me at the station in Cambridge. Barcroft introduced me to Dr. Benjamin Moore and to Sir Walter Fletcher, chairman of the British Medical Research Committee also director of the Natural History Department of the British Museum. Through Sir Walter I learned of the present work of Dr. C. S. Myers who had left the department of Psychology at Cambridge, and is now living in London practising Psychiatry and laboring for the establishment of an Industrial Institute for Applied Psychology in England. He gave me the London address of Dr. Myers.

Guy's Hospital Medical School, Physiological Laboratory,
Professor Pembrey and Dr. Ryffel.

Pembrey impresses one as having a very rugged constitution. He does a large amount of farm work as well as his physiological research and teaching. He is quite cordial and greatly enjoys exchanging views on current topics. At first he seemed a bit anxious as to whether his views gave me any offence. He said more than once, "We don't quarrel, we will be friends." He spoke several times of the heavy amount of teaching which was required with a depleted staff. His classes had to be divided into three sections of the same work. His research at present concerns graded exercises of physical work for heart cases. From a number of platforms he builds up stairways of different heights to provide physical exercise so the man can climb up one si side and down the other. As the case improves the stairway is heightened. "These cases do not improve in bed; you must work and get well."

Pembrey likes to research with men more than with animals. He told me of his work with marching soldiers. At four miles an hour in a breeze the soldiers would perspire, at three miles an hour they would not perspire. He used perspiring as a gauge of the limit of speed to employ. The soldiers were allowed to take off their jackets on the march. This never before had been done. He complained about the soldiers' uniforms and the public prejudice in favor of having the soldier always dressed up in a way which is not at all suitable for marching or for physical work. Pembrey was engaged in editing one volume of Luciani's Physiology, the English edition. The volume is on Metabolism and Reproduction. He said it was a poor translation in Italian and required much time in editing but he had promised to do it. It will be a large volume.

29

Sir George Dancer Thane, the British anatomist, died on Jan. 14, 1930 at the age of 79 years.
Science 14 Feb. 1930, p179.

Fig. 8 - Sir George Thane, Chief Inspector for the Anti-Vivisection Act in England taking tea with Professor Pembrey. He said the newspapers must not see this picture.

Dr. E. Mallenby of King's College for Women. Snap taken at Paris when at Physiological Congress.

Fig. 7 - Professor M. S. Pembrey in his office at the Physiological Laboratory at Guy's Hospital Medical School, London.

He is strongly averse to the popular operation of removing adenoids, tonsils, and appendix. He says they pay too well for the surgeon and it seems impossible to believe that the perfect physiological organism needs this assistance from the surgeon. A successful appendix operation according to him shortens the life by an average of six years. He laid special stress upon the fact that he had taken into account only successful appendix operations. He talked of the conversion of fat into carbohydrates, of reversible fermentation action, of the necessity according to his view, of one's doing his own calculation in order to use prpper discretion and to be thoroughly familiar with the work, of his belief that the LaboraParty in England is favorable towards science and not to be feared by scientific men. He declared himself firmly against birth control,"You never can tell when the brightest and most talented is to be born", stating that although it sounded cruel, that each family shoud have more children and let the best survive. In agricultural circles they had long ago learned not to breed from the first born. Pembrey has no sympathy for the man who overworks, i.e., has a nervous breakdown. He said certain physiologists, he would not tell the names,seem to rather pride themselves in their condition of nervous breakdowns. We discussed at some length the possible biological significance of the great over surplus of women due partly to the killing off of so many men and partly to the fact that infant mortality is greater in males than in females, the male babies being harder to deliver and also harder to rear. He believes that physiology indicates the proper sphere of man and of woman. They have certain physiological characteristics which fit them for certain parts of the cooperative work and endeavor of society.

In discussing the topic of the physiological effects of alcohol he said they had a very splendid opportunity to do research on the relation of the drink habit to infant mortality and diseases in malnutrition. In the hospital they have a large number of cases in which alcohol is a factor. Dr. Charles Cameron has charge of their child welfare work and they have an excellent opportunity for following up cases in connection with family habits, the children's health and death statistics, etc. He said they were very anxious to secure funds to proceed with this work and that a grant of two or three hundred pounds would enable him to appoint a young physician as a "fellow" to engage in this research. He said that although two members of their board of governors were brewers, they were very broad minded and allowed publication of any scientific material. I told him that last December Baldwin, director of the Child Welfare Institute of Iowa, had sought advice of me relative to using a fund of $50,000 supplied by the Women's Christian Temperance Union for work on the subject of "Alcohol and Children". I promised to write to Baldwin and lay the case which Pembrey had outlined before him as a favorable way in which to spend part of this money with a prospect of obtaining very useful and important results. We had lunch together with Dr. Laidlaw, lecturer in experimental pathology, Dr. Cameron in charge of Child Welfare Work, Mr. Hughes, assistant surgeon, (Note the use of Mr. and not Dr. in referring to a surgeon) and Dr. Poulton, assistant physician. Following lunch I saw, through the kindness of Dr. Poulton, the framework of the large chamber to be used for oxygen treatment in the hospital. This chamber is practically a duplicate of the one which Barcroft has in use at Cambridge. See figure . It was to be finished and ready for work in September, 1920. In the yard just in front of Pembrey's office they had been lately making some excavations and had come upon some graves of old victims from the date of the great plague in London.

Pembrey kindly showed me the anatomy museum in which the Towne wax models of human physiological preparation are specially noteworthy probably being among the best wax models that have ever been executed. I was especially impressed with a set of models which portrayed a large variety of skin diseases. The disections are also most remarkable and show that no pains have been spared to prefect every detail even in those places where the model could scarcely ever be examined by reason of the position in which it is placed or mounted on its baseboard. These models were made by one Robert

Towne who spent practically his whole life in this work at Guy's Hospital and the secret of his art he never communicated. He died in 1879. Pembrey showed me about the rooms of the department of physiology. There seemed a score of these rooms mostly quite untidy and with an unoccupied look. There were quantities of old dusty books everywhere. The rooms seemed deserted due to the fact that no work had been in progress during the war and the staff is very small. I saw his large balance with chair on each end, his big absorbing jars, and other pieces which are familiar to the workers of the Nutrition Laboratory. He had no new apparatus. His office is very nicely located, has beautiful tile walls and immense windows. It was here that I took the picture of Pembrey, see figure   . His desk space seemed to me decidedly limited. There were only two little desk tables piled high with letters and papers. Reprints were piled about on the floor in process of being sorted for distribution. (He will soon send this Laboratory a set of recent papers.) Above the fireplace which is near the entrance door of his office hangs the two tables from the "Publication 203, A Study of Prolonged Fasting," from the Nutrition Laboratory, and over the large table Pembrey has put up a little quotation to which he pointed when I was looking about his room. This quotation was from Milton, "Give me, above all other liberties, the liberty to know, to utter, and to argue freely, according to conscience."

In the afternoon Sir George Thane, see figure   , came on his tour of inspection of the Guy's Medical School and Physiological Laboratory. Hence, I had a second opportunity of meeting and conversing with this pleasant and broad minded man. Sir George remained for tea and was not averse to the camera. At the time of taking the photo, he said, "This picture must not be shown to the newspapers or given to them for many would think it quite disgraceful for the inspector under the Anti-vivisection Act to take tea with a physiologist." Tea time was very pleasant and we all became so interested in discussing the problem of under nutrition and the research carried out at the Nutrition Laboratory during 1917-18 that we allowed the tea-pot to melt apart!.

I had a short conference with Ryffel of the Department of Pathological Chemistry. He has not much teaching on hand but he is spending his time investigating cases in the wards and said he had little for publication. Complained of difficulty in getting things ready to publish because there is no clerical assistance. His chief interest at present is in gastric problems One assistant is working with him taking the Einhorn tube samples from patients every fifteen minutes after ingestion of food. He commented, "The idea that you can diagnose gastric ulcer from the shape of the curve from the results obtained with these successive samples is asking too much of nature." He was very much interested in what I could tell him of the work of Dr. Carpenter of the Nutrition Laboratory, on rectal feeding. He is also spending some time on problems concerning the kidneys.

London University, Imperial Institute, Physiological Laboratory.
Dr. Waller.

I visited Waller in his laboratory upon several occasions. One very seldom finds a man of his age (63 years I think. See figure 9), who exhibits in such marked degree zeal and enthusiasm for research work. At present his interests are along three particular lines. (1) plant physiology, the rates and conditions of plant growth, (2) the psycho-galvanic reflex, and (3) the production of carbon dioxide during the muscular work of man. When I came to his laboratory the first time (after some difficulty in finding the place) his son, Mr. Jack Waller, who is a botanist, and Mrs. Alice Waller, his wife, were working with him. On that particular afternoon they were working with some plant stems repeating research of Dr. Bose of Calcutta who works on the electrophysiology of plants. Waller is studying the growth changes of these plant stems uses a magnification of 1,000 to 10,000 times in place of the magnification of 1,000,000 to 5,000,000 employed by Bose. He said, and I think rightly, that 1,000,000 to 5,000,000 times seems a very unnecessary degree of magnification if the actual changes in the plant are significant. Weak teteanizing currents from a faradic coil cause the plant stem to longate while a strong teteanizing current causes it to contract. Bose interprets the former as growth. Waller thinks it is a matter of moisture change in the cells of the stem and not atlall growth. He claims that he can produce the same effect with a moistened fiddle string. For purpose of designation he prefers to work with fiddle string "E". They were taking photographs of the changes in the length of the stem using a small weight on the end of the stem to keep it stretched out and stimulating with two lengths of faradic current. It was most interesting to watch them experiment quite oblivious to their visitor. Waller dressed as for the street looking exactly like his photograph for the

Fig. 9 – Professor A. D. Waller at his desk in the Physiological Laboratory, Imperial Institute, London, writing up some experiments on the physiology of plant growth.

Fig. 13 – Dr. and Mrs Waller (and W. R. M.) in their garden studying the manner and rate of growth of the Lupine.

Fig. 12 – Gas analysis in the Waller Garden, 32 Grove End Road, London. Miss de Decker determining the volume of respired air in one bag. Mrs Waller writing the notes. Dr. Waller discussing how long to continue the walking experiment on the two young men shown in other views.

Fig. 11 – The pause for "Tea" in the walking experiment. Seventy rounds of the garden had been completed, thirty more were to follow. (Left to right, Miss de Decker, Subject, Mrs Waller, Subject, Peter, and Dr. Waller.)

last 30 years, with a pipe in his mouth, and a second pipe in his left hand was in charge of the stimulating faradic coil, Jack Waller attending the arc-light and the plant preparation, and Mrs. Waller looking after the camera. (This was a curious little affair made partly from a toy railroad track, an arrangement which they used in the early days in working with the electro-cardiograms.) Mrs. Waller was making the few notes, Dr. Waller puffing at his pipe, walking hurriedly back and forth across the room, making many chance and random remarks about the experiment.

It seems that he is in a hot debate over this matter of plant growth with the Hindu scientist, Bose. Waller worked on the physiology of plants particularly the action currents from stimulating plants some years ago and he believes that many of the results of Bose are quite spurious. I learned that the matter has been discussed between Waller and Bose by the exchange of several letters through the medium of the London Times. When Bose made his demonstration before the committee of experts at the Physiological Institute, University College, in the little room shown me by Bayliss, Bose refused admittance to Waller who was there at the time and desired to see the demonstration. Waller brought his own apparatus to the Institute that afternoon and demonstrated it to the same committee or to some members of this committee. This dispute between Waller and Bose is very loud and hot. Waller even thinks the latter an imposter, that he has been given a hearing in Great Britain and has been knighted largely for political reasons.

Waller has been doing the very profitable thing of studying plant growth not only in the laboratory but in his own garden. He laid off marks with India ink one centimeter apart on grasses and other varieties of plants and then measuring from day to day the amount of growth in the different segments marked and thus locating those portions of the plant stem which grow the most rapidly. He shows very easily that grasses grow at the bottom of the blade and not along the length of the blade or at the top. (See figure   .) Certain other plants when they have reached the blooming stage, e.g., the Lupen,

Fig 10 – Dr. and Mrs Waller with their assistant Miss de Decker conducting an experiment on the CO_2 production during walking. Note the rubber tube for the subject to step on at each round. This registered by a pointer on the kymograph.

My exposure was too slow.

Subject: Maw. age 35, weight 72 kilos, height 1.75 M; Surface = 1.88 M²

	Gross Cost in cc's CO₂ per sec	−Basal cc's/sec	=Nett cc's/sec	cc's CO₂ per KgM	KgM's per cc CO₂	Mechanical Efficiency (KgM/CO₂ ×440)
Bicycle ergometer at 30 steps/per min. (=35·26 KgM/per sec)	8·8	−3·8	=5·0 (700 Kalories per hour)	1·419	0·705	29 %
" " " 60 " " (=7·052 ")	15·67	−3·8	=11·87 (237 ")	1·684	0·594	24 %
" " " 80 " " (=9·128 ")	23·8	−3·8	=20·0 (400 ")	2·191	0·446	18 %
Staircase ascent (20 metres in 72·3 sec) (=20 KgM/p.sec) average of 10 ascents	27·5	−3·8	=23·7 (474 ")	1·185	0·844	34¼ %
Horizontal walk (6480 m. in 59'10") or 1·826 m./per sec i.e. just over 4 miles/per hour	28·8	−3·8	=25·0 (500 ")	0·190	5·250 (app)	—

Note † 4 miles per hour is rather above the normal walking pace of this subject, and he exhibits objection sign of this by his increasing cost of CO₂ during the hour: i.e. from 0·1484 cc per KgM at the start to 0·2115 cc per KgM at the finish.

The average cost is 0·19 as compared with 1·185cc for the true KgM. So that the Vert. ratio comes out only 6 instead of 10 which I think. Hor for the normal ratio 20 m ascent in 60 secs as a standard ratio for the normal ratio and 3·5 miles per hour walking speed. ADW

Image Not Reproduced

grow chiefly in the spike which carries the flowers. Waller maintains that Bose in some of his demonstrations used the portions of plant stems which would not normally grow in the parts employed.

On the second line of research, i.e., the study of the psycho-galvanic reflex, Waller has been interested for a long while. He seemed to welcome the opportunity (as did I) of using me as a subject in this experiment. A pair of electrodes were placed one on either side of the calf of the left leg. They were connected to a faradic coil to give a single induced shock as a stimulus to arouse "emotive" response. A pair of electrodes which seemed like large leather covered buttons, about 1½ inches in diameter, the leather moistened in salt solution, were placed one in the palm of the left hand and the other against the back of the hand, and secured in position by two rubber bands. This pair of electrodes through a suitable resistance box was connected with a slow moving coil galvanometer. Another pair of electrodes was placed on the forearm and connected to a second galvanometer. The light beams from both galvanometers were focused so as to fall upon the same photographic plate. I was sitting in a large well upholstered chair. After I had been quiet for some time, at a moment not exactly anticipated by myself, I felt a strong shock in my left leg. According to Waller, there was a marked response shown by the galvanometer connected to the palm and back of the hand and no response whatever in the galvanometer connected to the electrodes on the forearm. He said this was normal, i.e., a normal subject gave a reflex with the palm and not with the skin of the back of the hand or of the forearm. He continued, however, that certain subjects who had been tried by him were "palmer" also on the arm, i.e., they gave a reflex with both sets of electrodes. His observation up to that time led him to believe that the subjects who did this formed a peculiar group and he was inclined to think they were those individuals who are mediumistic or double personalities. I understood him to believe it possible that the psycho-galvanic reflex might become an important means of classifying individuals in regard to their mediumistic or psychical tendencies and powers. He said he felt the psycho-galvanic reflex distinctly the most important thing he had ever worked with. He thought that no matter if the galvanometer disturbance is due to the action of the sweat glands it was still an extremely interesting thing that this action of the sweat glands took place usually after such stimulation and only on the palms of the hands or the soles of the feet and not on the rest of the surface of the body. He claimed to have very carefully searched the surface of the body in reference to points from which to lead off for the reflex. Several people were named who had served for him as subjects, some of them very prominent men whom he had classified by this means.

Through a mal-adjustment of the camera he failed to get a photograph of the very first response in my case. Upon later trials with me it was evident that only the electrodes connected to the palm gave a response. I was interested also to notice that any increase or decrease of pressure on the electrode pad in the palm tended to send the beam of light to the right hand side which in this case was opposite to the direction which it moved at the time of emotive response. "This," said Waller, " is a very nice dispensation of Providence." This little experiment with different pressure on the electrodes I thought gave odd results. The galvanometer deflection at the time of emotive response was in a direction which indicated decrease of tissue resistance. Pressure would likewise be expected to decrease the resistance by reason of better contact and possibly some slight increase in area of contact. It may be that pressure on the wet leather caused it to become a poorer conductor. I expect the result was due to some peculiarity of the electrodes. Waller gave no hint that he understood the matter or had arranged to have it thus. There was a latency of about two seconds after the shock before the response occurred. Waller apparently knew nothing of any American literature on the subject.

He showed me many charts with the results of his CO_2 measurements on various kinds of muscular exercise. Most of these had been printed in short papers or communicated to the Royal Society. He told me of his work on the matter of stair climbing. The laboratory is on the fourth floor, and at the back of it there is an elevator. Around the elevator is a circular stairway. The climb on the stairway from bottom to top is 20 meters. He can have the subjects climb up the stairway and then use the lift to take them down to the bottom from which they start and climb up again. In one of his experiments he used four men for subjects. They climbed the stairway 25 times the first hour, 25 times the second hour, 20 times the third hour, and 10 times the fourth hour. This made all in all considering the weight of the men 100,000 kilogram meters of work for each subject. The men were so nearly exhausted at the end of the four hours that he considered this was too much work to expect a man to do in four hours. These men were in a trained condition and physically fit soldiers. Waller considered that 100,000 kilogram meters is not too much to expect of a trained man during 8 hours of work. He thinks that a man can produce about 1/20 of a horsepower throughout the day and on forced marches can do the work of 1/10 horsepower. His results are in published papers and need not be given in more detail here.

Waller is fond of using his visitors as subjects in the experiments that he has on hand. His methods may be characterized as of a "rough and ready" or practical sort which allow him to get such results as he wants without great refinement either in technique or calculations of data. In his muscular work researches he considers only the CO_2 production and analyzes only for that. He collects the expired air in a rubber bag or "pillow" as he prefers to call it, collecting the air for about $\frac{1}{2}$ minute at stated intervals throughout the work. A nose-clip is in place during the period when the air is collected in the pillow. Other English physiologists working with respiration

physiology object seriously to this rough method. For example, Cathcart of Glasgow said, "Oh, Waller is out heartbreak. I wish Benedict were here to talk with him like a brother and explain to him his errors". However, Krogh

Benedict, and others believe that Waller is using a method which although rough, does bring out the large differences in the CO_2 production of the exercise with which he is dealing and he does have just enough ground on which to stand and put up a good fight. Waller is rather fond of arranging a terminology of his own and of using all sorts of short cuts in computation. I served as a subject on the bicycle ergometer (Martin type), riding it for five minute periods at different tempos. I climbed the stairway 10 consecutive times about as rapidly as I could do it, and another day, walked around Dr. Waller's garden 30 times. The distance around the garden is   meters. Some idea of the outdoor experiment in muscular work which can hardly be termed "the common or garden" variety may be gained from the photographs. See figs.

Lecture at University of London Club.
Dr. J. C. Bose.

Aveling informed me that Bose was to lecture at the University of London Club. I secured an invitation from its secretary, Mr. Humberstone, but was a little late in finding the place. The room was well filled and many were standing. Bose was describing the conditions which must be met in building apparatus to use in registering the growth movements of plants. The friction of a tracing lever in smoke on a kymograph drum was entirely too great to be overcome by the strength of the growing stem. He described an arrangement of intermittent contact of the recording point with the smoke which left the lever free from the friction of the smoked surface. Then he described what he considered his best instrument for registering plant movements. This he called the "cescograph". He showed a diagram of the arrangement. The apparatus was present but it was housed in such a way as to make it impossible to carefully observe it. The principle of the apparatus is to use a magnetized lever point which moves past a magnetized suspension. Therefore in place of getting actual contact of the tracing lever with any surface he gets a magnetic linking. It appears probable that the movements of the magnetized suspension are not in a linear proportion to the movement of the magnetized lever but as I recall, nothing was explained about these factors or characteristics of the instrument. A small movement, elongation, or contraction of the plant stem causes a relatively very large movement of the magnetized suspension, which, carrying a mirror and reflecting a beam of light, is further magnified a great many times so that the result in magnification is some one to five million times the change in the plant stem length.

Bose showed many lantern slides most of them reproductions of kymograph records in which his intermittent recorder had been employed. By these slides he proposed to demonstrate that plants respond to stimulae, that they (certain of them) sleep, that they have connective tissue which corresponds to the nerves of animal organism, that these plants show action currents as do the muscles and nerves of animals, that the plants or certain of them have movements, som of them of a pulsating character, e.g. the plant known as Desmondrum, native to India. He demonstrated also the growth curves and the conditions which alter these such as sunlight, fresh air, and electrical and chemical stimulae. He also showed records indicating the death spasm of the plant and tried to demonstrate with his cescograph a plant in death spasm. He had of course to interpret the meaning of the moving beam of light on the screen and the spectators had to believe his explanation. His lecture was quite oratorical and there were many things put in to please the British audience. I saw clearly how the subject was of special interest to a Hindu with his characteristic views of life such, e.g., as the transmigration of souls. Bose said at one place, "Therefore there are not two streams of light that never meet but life is one. There is only one stream." He also said, "I have had to wait a long time for recognition of my work and am sorry that not all are even now satisfied." Mr. Jack Waller was present and took notes.

Note. Professor Sir Jagadis Chunder Bose, Kt., F.R.S., M.A., D.Sc., LL.D., C.S.d., C.d.E., is founder and director of the Bose Research Institute, Calcutta, India. His complete lecture on "Plant and Animal Response" is published in Proceedings of the Royal Society of Medicine, June 1920, Vol. XIII., No. 8, pp. 101-128. This contains also the discussion which followed the presentation of the paper, a discussion chiefly between Waller and Bose.

King's College for Women,

Dr. Mallenby.

I had the pleasure of visiting twice with Mallenby at King's College for Women. At one of these times I heard him deliver a lecture to his students and some visitors on the subject of "Food requirements for Basal Metabolism and for Work." He referred considerably to the research of Benedict and Cathcart. Another meeting which I remember with much pleasure was when Mallenby and myself together with Dr. Dale and Dr. Cullis rode from Cambridge to London together after the meeting of the Physiological Society and discussed alcohol experimentation and physical research. Dr. and Mrs. Mallenby are working on the subject of rickets having a number of dogs which they use as research animals. These are in Cambridge. This work was reported at the Paris Congress at which place also I saw something of Mallenby. See figure 57

Mallenby has been interested in preparing a food with the anti-ricketic substance in it to take the place of cod liver oil which he found to be the best food with which to combat rickets. He made or caused to be made a mixture of beef, suet, peanut oil, and glucose which substance they called "Butol." Quite a quantity (3 barrels) of this was sent to Vienna to be used as human food.

Mallenby has continued his work on the subject of alcohol and has some unpublished research of this nature. He worked with one human subject, gave him various amounts of alcohol even to the point of making him thoroughly intoxicated. He determined the amount of alcohol in his blood. The only test for the man's voluntary efficiency in control of muscles was to have him draw a circle and two diagonals through the circle without any aid or pattern. I saw many of the drawings but no tables of data. The drawings showed much decrease in accuracy.

The college building is new and Mallenby had very fine light laboratory rooms but he is to leave the position for another this Fall.

The professor of Chemistry, Dr.   , at the King's College for Women is interested in the problem of vitamines from just the chemical standpoint. He had what I judged to be a very good quartz prism arrangement for measuring the spectrums of different solutions containing varying quantities of the vitamine substances. He kindly showed me the general arrangement of the building which was indeed, light, convenient, and attractive, with an especially good roof garden overlooking a large section of the city.

Ministry of Health, Whitehall, London.
Sir George Newman.

We had much to say of mutual friends both in America and England. I explained the nature of my foreign visit and commented upon the important group of physiologists who are now living and working in England. Sir George being the chief medical inspector of England is of course acquainted with all of these men and it is his duty frequently to investigate their work and teaching as related to medical education in Great Britain. He spoke of Sherrington and Starling as the best of the present English group, favoring the former if he had to have the most expert opinion for government purposes as sometimes he found needed. Schafer he felt had passed his prime. Sir George spoke in highest praise of Hopkins and gave me a card of introduction to him. Concerning L. Hill he said, "He may be good but he is a propagandist of so many kinds and goes about the town helping this and that cause, all very good, but not the way to be a great scientist. You must be right at your job."

We discussed the Liquor Control Board of England and the book recently published by them entitled "Alcohol, Its Action on the Human Organism". Sir George informed me that it was his duty to select this sub-committee and assign the work. That the book had been written by Dale and Greenwood. He expressed much interest in any alcohol research of the Nutrition Laboratory. He spoke highly of McDougall. Sir George referred to their experience during the war in cutting off alcohol and increasing the food and thereby greatly increasing the physical output in munition. We discussed somewhat the psychological and medical work done on the problems of aviation during the war. I also received advice as to individuals whom it might be worth while to visit. While my interview was rather short because of the heavy administrative duties of this man, it was a great pleasure to come into contact with this strong personality. Newman's reports on Medical Education in England give current reviews of British Physiology.

National Institute for Medical Research, Mount Vernon,

(British Medical Research Committee)

Dr. L. Hill.

Dr. Hill assisted in his laboratory by Miss Harwood-Ash, (See fig. 14) showed me his "Kata"-theremometer[1] and had much to say about it as a convenient instrument for measuring the cooling power of the air under different conditions of ventilation. He is especially interested in the problem of ventilation and outdoor air treatment and had published recently, as one of the Medical Research Committee's publications, a monograph which was to be the first part of his contribution to this field. He gave me a copy of this monograph. He is convinced that the <u>comfort</u> of physiological organism working in a room or out-of-doors depends on the cooling power of the air. With Miss Ash he was trying to develop a "Kata" as he continually called it, which as heated electrically by a small resistance wire imbedded in the bulb. He heats the theremometer to a certain temperature and then notes the time elapsing while it cools a certain number of degrees. He spoke much about the results obtained in wind tunnels and said they do not agree with results obtained on top of the roof with the wind blowing at the same rate. He thought that the radiation from the surface of the ground probably effected the cooling power of the air. To my question if he had performed the experiment out-of-doors at night, he answered that he had made no night experiments and thanked me for the suggestion.

Hill hopes to become connected with various factories in the capacity of consultant from the Medical Research Committee and to measure the fitness for work of the employees under different factory conditions. He said he would use the "Flack Methods", i.e., measure the vital capacity, the length of time that the subject

---

[1] The Kata-thermometer may be purchased from James Hicks Company, 8 Hatton Garden, E.C., London, or the Sebie-Gormon agent in the United States.

Fig. 14 – Dr. Leonard Hill and his assistant Miss D. Harwood-Ash working with the kata-thermometer at the National Institute for Medical Research, London.

Fig. 15 – Dr. H. H. Dale's laboratory for Bio-chemistry and Pharmacology at the National Institute for Medical Research, London.

can by blowing hold a column of mecury to a certain pressure, etc., etc. Dr. Flack was away at the time of my visit. Hill thought that physiological tests which involved the tone and fitness of the abdominal wall were of the greatest importance. He referred to the "fundamental nature" of his early work on the abdominal wall and the blood supply and said that it was never cited in current physiological text books. He thought it of prime importance in considering physiological fitness.

Hill favors developing resistance in the physiological organism rather than trying to get rid of all carriers of disease which latter is the aim of Dr. Benjamin Moore. Hill criticises the work of the Nutrition Laboratory that all of our metabolism measures have been made in still air. He believes that the metabolic level is raised especially by cold and by wind and this not because of any shivering of the patient or any physical exercise of the patient due to his being in the cold or wind. He thinks this change in the metabolic level to be the benefit of the outdoor life. His hypothesis is worthy of consideration since it is known that outdoor life is favorable to health, to recuperation from tuberculosis, and such conditions, and since it is further known that the oxygen and purity of the air is probably not better outdoors than indoors. He thought the skin temperature of the cheek a good indication of the metabolism and said that by using a large bulb thermometer, moving and turning it a good deal would give a very satisfactory reading of the skin temperature.

I think Hill feels disappointment or lack of appreciation of his scientific work. He referred to his 20 years of teaching at the University of London and how it was necessary continually to "cram" men for examination to get them through so that they will in time be the "Consultants" of the West End. These 'Consultants" give their services free to the hospitals to make friends of the young doctors who will later call them as consultants. It is a vicious system according to Hill. The consultants belong to the aristocraticclass, they live off the aristocracy, and know only about the ills of the aristocracy, never seeing the inside of a factory, or knowing of the ills of the common laboring class. Medical education is not right according to Hill. Physics, chemistry, and biology should be taught in their application to medicine and not by so many separate specialists. Anatomy he thinks is a clear waste of time. The anatomists should be made to teach histology and bacteriology. Hill believes and practices the outdoor life and thinks you should be in the country and that part of your time should be spent in farming or driving. He is tall and apparently of strong physique.

This building which the Medical Research Committee is occupying was previously a tuberculosis hospital. They have just moved into the place. It was ready for them in 1914 but was taken over as an aviation hospital and the Medical Research Committee was given temporary quarters at the Lister Institute. The plumbers were to have the building ready in December of 1919 but the job was prolonged until March of 1920 and at three times the estimated cost. They are to have their own cafe on the top story of the Institute. About 60 employees will coöperate. It is a voluntary arrangement and aside from the funds of the Institute. The lunch for doctors will be at the rate of one shilling and eight pence. The assistants will pay somewhat less. There are to be a few beds so that men who have research on hand which requires their being present and looking at their material at night can stay there at such times. Also I was told if they have scientific visitors they could occasionally put them up. The coöperation of the Institute with factories is entirely voluntary and the investigators have no teaching duties. In a good many ways the Institute resembles departments of the Carnegie Institution.

Dr. Benjamin Moore.

Dr. Moore, the physiological-chemist of the Institute, was studying photosynthesis. His research might be characterized as on the problem of the first origin of the organic from the inorganic. The buds of spring he believes are not holdovers from the life and energy of the past summer but the result of the manufactured substance of late winter while everthing is still brown and apparently lifeless. The brown bark of the trees and shrubs is quite translucent at the points where growth is going to occur and act as light filters to the chlorophyll below cutting out especially the blue and ultra-violet light. Moore was measuring the spectrums of many samples of bark from tree limbs and shrubs. A second research problem that was interesting him was that of nitrogen fixation. He was growing twigs in different kinds of water and light and measuring the amount of nitrogen fixation by the amount of change in weight. He has a very convenient outdoor room associated with his laboratory.

Dr. Dale.

Dr. Dale's department of Bio-chemistry and Pharmacology was quite completely equipped and installed (See fig. 15). On a rigid beam which ran lengthwise in his main laboratory room and can be seen at the top he had many outlets for gas, water, electricity, and time from a Brodie clock. The beam was high enough to offer no obstruction to workers in the room and tables and kymograph supports could easily be manipulated below it, all wires and rubber tubing extending upward rather than down and about the floor. Below the beam and parallel to it was an opening in the floor for all liquid waste. The laboratory had two relatively new C.F.Palmer and Company's kymographs (fig.15) which included the power table and on this was mounted the artificial respiration pump. A little special room with suction ventilation was arranged for smoking paper and for dipping and drying the records,- a very complete installation. He had also a new long paper kymograph with two very small drums at one end. Between these two drums is a flat and relatively firm but large area on which the tracings of large, curved lever movements may be registered. The Palmer kymograph has signal magnets and manometer mounted on it as a part of its regular equipment.

I did not learn of the researches that Dr. Dale had in progress. There was at the same time another visitor, Dr. Alonzo Taylor, of America, whom I was very glad to meet here. He was just returning from an extended visit through Germany and Austria in reference to the conditions of nutrition in those countries. Taylor urged me to visit Vienna and see the work of Dr. Von Pirquet in charge of the childrens' clinic and the Hoover Food Kitchens.

Dr. Brownlee.

I was shown the excellent equipment of the department of Statistics by Dr. E. Schuster secretary of the publication department and Mr. Russel of the Statistics department. They had one of the "Millionaire" computing machines and said it did addition, (non-listing), and multiplication for which they considered it the best existing machine. It requires one turn for each place in the result. It does division quite well, but subtraction quite slowly. It is a large, heavy, box-like devise. They had the Euclid computing machines both electric and hand operated and said it was especially good for division. There was a large ruling and plotting machine in a special alcove and above this machine a large aluminum cover nicely counterpoised so that the machine would be kept out of dust whenever not in use. It was a Swift machine. A second curve drawing instrument was especially adapted to draw a given curve to a different ordinate value or to change one dimension of a picture or figure. There were two harmonic analyzers both designed and one made entirely by Schuster. It was a beautiful piece of work. I believe Brownlee is devoting himself to medical and vital statistics. His was not in any case a department to give statistical treatment to the data collected by the other men working in the Institute.

Dr. Schuster

Dr. Schuster as secretary of the publication department edits and sees reports through the stages of proof and final printing and has charge of mailing and all applications for the reports of the Institute. Schuster was the first Galton Fellow. The Eugenics Laboratory at University College now headed by Dr. Karl Pearson is the development of that fellowship. He has worked with Dr. McDougall at Oxford in the department of psychology, and arranged a modification of McDougall's Dotting-Machine. He therefore was interested in such problems as testing men for special abilities, e.g., aviation. He gave me the reports of the Medical Research Committee on the subject of aviation and the testing of aviators and also kindly entered the name of the Nutrition Laboratory on their mailing list for all publications.

I was especially anxious to learn about the methods and policies of the Institute. Schuster gave me the set of annual reports and discussed the early organization of the British Medical Research Committee which now is under the Privy Council. They have no endowment. The budget is a matter of an annual grant. Its amount and the expenditures of the committees are not made public. Each research man is his own boss so far as problems and work are concerned. There is a stock room for the more general supplies upon which they can all draw but for any special apparatus or for the undertaking of any special problem a man must himself go with a requisition direct to the Board which means to Sir Walter Fletcher. The Board is quite liberal and willing to listen to propositions for research.

In reference to their methods of getting out publications we discussed the dictaphone and the typewriter neither of which they employ. They take the original long-hand copy of the author. He, however, may if he desires, use a typewriter, but they in general have no secretaries or typing assistants. Schuster goes over this copy and makes notes giving directions for the type-setting and sends it in for composition. He thinks there is less correcting necessary and a greater saving of time to the author with fewer mistakes when you use the long-hand copy of the author but that it requires a better grade of compositor and so on this side costs more. He showed me the fearful condition in which some of the copy comes to him. He said it requires about seven months after a monograph has been sent to the printer before it is completed. He thought most of the delay was due to the time the author himself took in reading the proof. There is one report on hand, that of the Wassermann Commission, which has been on the way two years but this is because of extensive revision, the committee having changed its mind in reference to several points. As an institute they employ no draftsmen. The printer has the author's sketches redrawn. Schuster himself frequently fixes up a sketch first if the author does not choose to try to make one. Very many of their reports come from people outside the Institute, so he thinks it would be difficult to have a draftsman in their employ. They turn out a large number of relatively short monographs with paper or cardboard covers and

Fig. 16 – Dr. Edgar Schuster in his private machine shop at his residence, 110 Banbury Road, Oxford.

Fig. 17 – Dr. Schuster, Secretary of the Publication Department of the National Institute for Medical Research, London.

Fig. 32. A circuit interrupter for use in investigation on nerve physiology. Made by Dr. Lucas but not described until 1921 by Adrian.

get them out, it seems to me, very promptly.

As a hobby, Schuster engages in machine work. He has a small machine shop behind his home in Oxford with a very nice equipment of lathes, planer, etc. The milling machine he himself designed and built (See figs. 16 and 17). I have previously noted that he designed both and built one of the harmonic analyzers for Brownlee's department. At the time of my delightful visit with him at his home in Oxford, he was building a bacteria grinding machine for use in the work of the Institute. His interests and capacities are as varied as they are noteworthy.

If the National Institute for Medical Research now located at Mount Vernon, Hampstead, continues to receive the support of the British government, it will probably be one of the most important research agencies in England. Its field is very wide and one wonders if it will gradually come to absorb such organizations as the British Fatigue Board which is interesting Sherrington and also Lee of Columbia University in U.S.A. Dr. Charles Meyers is trying to establish an Institute for Applied Psychology and Physiology. Sir James McKenzie is endeavoring to establish an Institute for Preventive Medicine, the object being to develop tests and diagnostic methods which can be used to determine the tendency toward certain diseases before they have developed in the individual. It seems to the outsider like the National Institute for Medical Research and might well be the organization through which all of these lines of research and application of science to medical practice might be carried out rather than having several institutes with overlapping fields.

However, among the names of supporters for the proposed new institute we may note those of Sir Walter M. Fletcher, Sir R. A. Gregory, Mr. W. B. Hardy, Dr. Leonard Hill, Sir Alfred Keogh, Dr. C. S. Myers, Sir E. Cooper Perry, Prof. C. S. Sherrington, and Prof. E.H. Starling. The intention of the founders is to estalish a national institute which will investigate the human problems of industry and commerce, occupying a position similar to that held in the domain of physical science by the National Physical Laboratory. It will provide training courses and lectures for those interested in the practical applications of psychology and physiology to the problems of industry and commerce. It will undertake investigations at factories and offices in relation to any special problems. The institute will not be established for profit, and a close relation will be maintained with the Industrial Fatigue Research Board, but overlapping of effort is to be avoided.

British Psychological Association, Bedford College, Regents Park.
Dr. Scripture and Major Kline.

This meeting was held on Saturday, May 8th. I attended at the invitation of Spearman who was in the chair. The first communication was by Dr. E.W. Scripture on the subject of "Speech Inscriptions in Normal and Abnormal Conditions". The main contention of the address which was well delivered was that the voice by its inflections which are matters of pitch, expresses most fundamentally and delicately the psychological states and conditions and that these cannot be masked or misrepresented. The voice is a direct and certain indication of the psychological condition in the subject's mind. He claimed that as a means of diagnosis in epilepsy the speech curves of the voice have no equal. The epileptic does not modulate his voice as the normal individual and although the ear cannot discover this without training, the epileptic voice shows really but small pitch differences with different words of a sentence. This is an expression of his unyielding attitude to the world. An epileptic fit is a temporary lapse to ease up from this struggle of unyielding resistance to outside conditions. The epileptic is not adaptable. This voice sign is a very early diagnostic. In the army it proved the reality or sham of an epileptic fit, but the author claimed the epileptic could be trained to evade this test. Spearman comented upon the paper by saying that when you are playing a card game he had noted you can control your face much better than you can control your voice. Following this address the meeting adjourned for tea.

I was especially glad to hear Scripture. He was once Professor of Psychology at Yale University and among the foremost in this country who were devoting themselves to experimental psychology. It was under him that my good friend and former professor, Dean Carl E. Seashore, served as assistant and received his early training in psychology. Scripture has for a long time been interested in voice curves and the Carnegie Institution has given support to his researches which have been reported in year books of the Carnegie Institution, Nos. 2, 3, 4, and 5, and in the Carnegie Institution Publication No. 44. Scripture is now practicing medicine as a "consultant" in London. At tea time I was able also to meet and talk with Professor Cavenish Read formerly Professor of Psychology at University College. Following "tea" Major Kline with many pictures and much lucid discussion showed what kind of camouflage was effective and what was worthless from the standpoint of land, sea, and air warfare. The paper raised a good many problems in vision and space perceptions.

University of Edinburgh, Physiological Laboratory.
Professor Edward Sharpey Schafer and Dr. W. W. Taylor.

Professor Schafer was at his home recovering from a recent illness at the time of my visit. (See figs. 18 and 19). He was most kind in inviting me to see him at his beautiful residence in North Berwick. After inquiring about the Nutrition Laboratory and its personnel he informed me of recent developments in his own department in Edinburgh. It seems that Dr. Cramer and Dr. Bruce who were in his department at the beginning of the war were both Germans or of German extraction. The University authorities asked them to resign as a matter of policy but informed them that they would be reinstated. Bruce resigned but was not reinstated. Cramer did not resign but insisted they had no right to require his resignation under these circumstances and it seems they could not discharge him. However, he accepted a position in London with The Imperial Cancer Research Commission. Dr. John Tait, also a member of the department, became professor at McGill University, Canada. With these three men leaving, Schafer was practically alone in his department. Cramer's place has been filled by Dr. W. W. Taylor, a chemist, and the Histology is now taught by a very bright Japanese, Dr. Lim, and a lady, Dr. May Walker.

Schafer admitted that the teaching work was quite heavy but he thought there was left plenty of time for research work if the man or woman really had a keen desire to do this. He had considerable to say about originality and research, i. e. the ability to conceive and work out problems in an original way. He thought this quality was a great feature to be sought in men who were to become members of a department in physiology or in any of the medical sciences and that it was very rarely found among women.

Schafer said that he was continually fighting with the medical faculty in Edinburgh. He himself wants to erect a new physiological laboratory on a tract

Fig.—60 - Prof. Schafer and family, host of
1923 Congress at Edinburgh.
See opposite p 124.

Edinburgh, Albert's Seat from North Bridge.

Sir James 'putting'.

Fig. 19 - Professor Schafer was convalescent and had not been out of doors for several days.

Fig. 18 - Professor Sir Edward Sharpey-Schafer at his residence, North Berwick, Scotland.

of land which they have bought outside Edinburgh. They insist that all the medical subjects must be kept together in the same building.

Dr. Taylor was kind enough to show me around the physiological laboratory and also to some extent the environs of the city. He was working on the problem of the acidity of urine and showed me platinum electrodes of his own make for getting the hydrogen-ion concentration. (See fig 20). He said that with these electrodes it was very easy to get saturation. He was employing a capillary electrometer, a circular bridge, and measuring the conductivity of electrolytes. He referred to two workers, McGreggor and Fitzgerald, who had used a rotating commutator arrangement in their measuring bridges. Taylor has made what he thinks is a much more satisfactory rotating commutator for such use. In his Wheatstone Bridge he uses a split-box in place of a slide wire. The commutator interrupts direct current for the source in his Wheatstone bridge and after this interrupted current passes through the electrolytic cell, the resistances, etc., it is collected by the other part of the commutator into a continuous direct current again which has no interruptions. A zero center scale galvanometer is used as an indicator in place of the usual telephone. Taylor expects to publish this work soon.

They had what Taylor considered a very wonderful balance made by H. Collot of Paris. The balance had the most simple bearings, would weigh one kilogram, and could be read to the nearest 0.0001 of a gram. You have but to put on weight to as near as 0.05 of a gram and allow one minute. There is a vane on an oil chamber at the bottom which provides damping. The scale is calibrated plus and minus and thus the pointer has not to come actually to a zero position. It has a microscope reading device. The small weights are bits of rod bent into three planes which make them easy to manipulate.

The large room in which physiology is taught is well fitted and convenient Schafer has a small book which the students purchase and during their course cut up and paste different parts of it in their notes together with the curves which they get from their practical exercises. They pay much attention to the history of Experimental Physiology. In suitable cases at the sides of the room Schafer has labeled many old and important physiological instruments giving the name, date, when used, and the name of the inventor or maker. At examination time a selection is made among these instruments. They are put out on exhibit without the name cards and must be properly designated as a part of the written examination.

I should greatly have enjoyed being at the physiological laboratory at a time when Schafer was there and feeling fully well and when he would feel like many comments of a physiological nature, stimulated by the presence of apparatus pictures, etc.

University of Edinburgh, Royal Infirmary.
Dr. Meakins.

Dr. Meakins, (See fig. 21), aside from his regular hospital work in the Infirmary, was trying to develop a method of oxygen treatment for pneumonia cases. He seemed quite interested in the Benedict Portable Apparatus and showed me another apparatus recently devised by Haldane, Meakins, and Priestly,(See their paper entitled, "The Effects of Shallow Breathing" British Journal of Physiology, Volume 52, Number 6, May 20, 1919). A description of the apparatus together with its use is given in this paper. They call the device "Concertina" apparatus for continuous record of respiration (See fig. 22). They use a mouthpiece and nose clip. The composition of the gas breathed by the subject is determined and mixed in a glass chamber (A). The expired air passes out into the room. The depth of respiration is controlable by the limiting position (See clamp B) for the bottom end of the bellows. From the movements of the bottom part of the bellows, C, a graphic record of the respiration is easily obtainable. Meakins had just received from Siebe-Gorman Company two kymographs, D and E, shown in the view. They are exact duplicates of those made by the Harvard Apparatus Company. I understood Meakins was unable to secure any kymographs from the latter concern.

While I was present with Meakins he was doing some work with blood samples taken from patients that morning. He was measuring the amount of oxygen these samples could absorb and then determining the amount of oxygen obtainable from the sample and in this way obtaining the amount of oxygen already in the blood when he began to work with it. This blood gas apparatus (See fig. 21) is made by Haldane. Meakins was speaking with Dr. Cushny and myself at lunch time of a patient in which he was especially interested. This patient that morning had exhibited both apnea and hyperpnia.

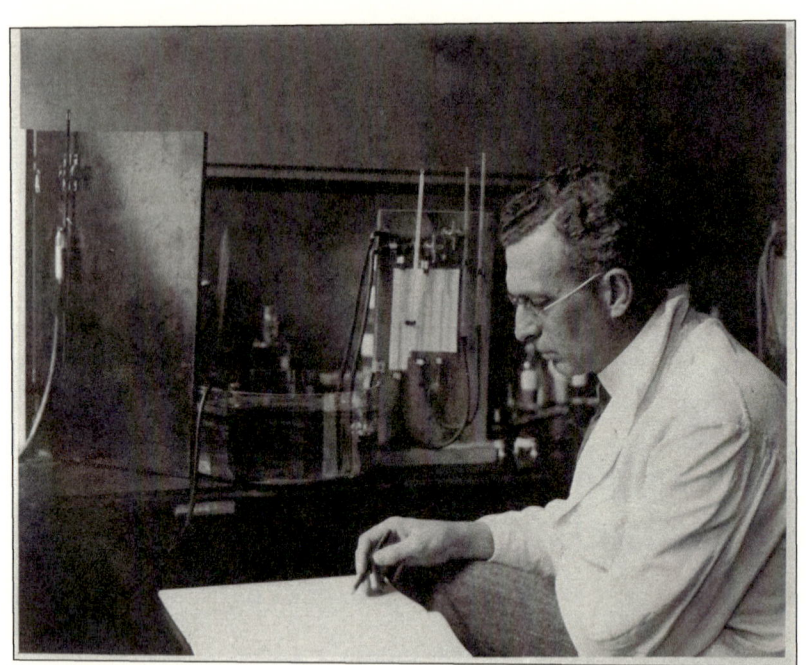

Fig. 21 – Dr. J. C. Meakins at the Royal Infirmary working on blood samples.

Fig. 22 – The "Concertina" apparatus for continuous record of respiration. Used by Haldane, Meakins and Priestley. See Jour. Physiol. 1919, 52, p 433.

Fig. 20 – Dr. W. W. Taylor in his laboratory for Physiological Chemistry, University of Edinburgh. The apparatus is for determining hydrogen ion concentration.

University of Edinburgh, Laboratory of Pharmacology.
Dr. Cushny.

Dr. Cushny (See fig. 23) has lately come here from University College, London The previous occupant of the chair of pharmecology at Edinburgh was not so much of an experimentalist so that Cushny has an opportunity of developing the laboratory and of fitting it out with an almost entirely new equipment. He showed me a very fine museum of natural drugs, poison arrows, etc., which the previous professor of pharmacology had spent much time in accumulating and classifying. Cushny raised the question of how the primitive man could have discovered the use of curara which is not poison when taken through the mouth but only when introduced into the blood and hence the natives dipped their arrow tips in curara. This museum he had moved to the top floor and the room made vacant by its removal had been converted into a splendid experimental room fitted up much in the same way as the new laboratory of Dale at the National Institute for Medical Research. Cushny also had arranged an overhead distribution for electricity, time, air, gas, water, artificial respiration, etc., running across the room at such a height that the experimenter could reach it but still not be bothered by it when walking under.

We discussed the work of the Liquor Control Board in England and its committee which recently put out the small book entitled "Alcohol, Its Action on the Human Organism". Cushny was a member of this sub-committee which prepared the above-named report and was interested in some account of the recent work on alcohol that had been done at the Nutrition Laboratory. At lunch time we were also with Meakins, and Cushny mentioned what he termed a "peculiar look in men's eyes after they take alcohol". We tried to analyze this, Meakins affirming that it was by the eyes one could tell that an associate had been drinking alcohol too freely. No one of us could be sure whether it was a peculiar stare of the eyes or a dryness of the eyes or what I thought quite possible, that the number of fixation points in a given amount of time is smaller. Of this I am not certain but since the eye movements are made slower, it seems possible that the eye will not tend to move so much, i.e. so continually change its fixation point, in the intoxicated man. The question is an interesting one and I think worth an attempt at experimental analysis since it seems to be a matter of quite common observation.

Fig. 23— Professor U. [illegible] in his newly equipped laboratory for Pharmacology at the University of Edinburgh.

Fig. 24— Professor E. P. Cathcart looking over news from the Nutrition Laboratory at his desk in the Physiological Laboratory, Glasgow.

Image Missing from Original

Fig. 26 – Dr. J. Barcroft's large, three compartment respiration chamber made with glass walls. Mr. H. Secker, assistant, standing in middle compartments, in view below. Taken with camera in compartment A. The CO_2 absorber shown at the right of the bicycle ergometer.

Image Missing from Original

Fig 25 – View taken from compartment C. One of the (F) food openings shown in the lower left corner. Mr. Secker Barcroft's personal laboratory assistant in the view.

University of Edinburgh, Psychological Laboratory.

Professor Drever.

Dr. Drever until recently has been teaching the subject of Education. At the death of Dr. Smith, who had charge of the Department of Psychology, Drever took up that work for the remainder of the year and now has been placed in charge of the Psychological Laboratory. His chief psychological interest up to the present has been along lines of social psychology. He has some points of view at variance with those of Dr. McDougall. Concerning some of these he told me and said he was going to presents his viewpoints at the next meeting of the Psychological Association hoping that McDougall would be there so that they might discuss them out. Drever expressed his desire to secure some technique which he could use to measure qualitatively whether certain psychological situations were emotive or not emotive. We considered the psycho-galvanic reflex as a possibility along this line. He had no desire to use it quantitatively.

The Psychological Laboratory here is the best equipped of any that I saw in England, I mean so far as quantity of nice apparatus is concerned. It seemed to me they had practically everything that Spindler and Hoyer have made. I saw a special pencil arranged with a tambour at the top so as to measure the pressure exerted during writing. It was a device of Drever's.

Drever told me he thought Psychology had come to its own as a result of the war. He said the Psychological Laboratory in Edinburgh was really established by George Combe who was a phrenologist, and his brother, a physiologist. It seems that George Combe died about 1850. He was at one time a candidate for the chair of logic in the University of Edinburgh against Sir William Hamilton. He was quite a prominent man, he made a fortune by the practice of phrenology, and married a daughter of Mrs. Siddon, of art gallery fame. The Psychological Laboratory can get practically what it wants in the nature of equipment for special purposes by asking the trustees of the Combe estate. Drever will probably do work of an experimental nature on social psychology. He thought that it was a great pity Oxford allowed Professor McDougall to be taken to America.

University of Glasgow, Physiological Laboratory.

Professor Paton and Dr. Cathcart.

It was especially interesting to visit with Cathcart since he has been a research associate at the Nutrition Laboratory and coöperated in the Carnegie Publication No. 187. He knew many of the present members of the Laboratory staff and was thoroughly interested in our situation and in looking at photographs from the Nutrition Laboratory (See fig. 24). Cathcart complained about the great many students who, as others have stated, held back by the war, have now crowded into the Medical School. He came to Glasgow with the hope that there would be opportunity for research and now he finds that owing to the poor health of Dr. Paton, he himself has to undertake much of the administrative burden of the department, also he has been made dean of his college. They have 612 students in physiology this year, as I remember, and 29 sections in the course of practical physiology. He has a certain amount of hold-over work from his position as instructor in the defensive gas warfare. Cathcart told me quite a little about his war experience and slight amount concerning the metabolism work which he did during that time. He said he published everything except some work that was done outdoors which gave quite unusual results. I referred to the contention of L. Hill that the metabolism was raised outdoors. Cathcart seemed to think this quite possible.

At first, thinking that my interest was chiefly in psychology, he professed to have nothing to show me. He exhibited quite an interest in psychical research and in certain psychological problems as for example, one's mental imagery at the time of writing scientific papers. He gave me a copy of a little elementary treatise in physiology which he wrote for the instruction of soldiers. He said that he wrote the whole thing at one sitting, that he kept his imaginary audience before him, and in reference to one part of this book he said, "I could see my audience sit up and take notice, when I gave them that part." Cathcart was greatly impressed with the work of certain men in Edinburgh who received the raw ewcruits for the army and gave these men their preliminary instructions and training. He thought he had never before understood the value of psychology or realized its sphere of influence until coming into contact with this case of where raw boys were very shortly by the instruction of these men made into genuine soldiers with a desire to fight.

Our conversation was so interesting that Cathcart even forgot "tea" where we would have had an opportunity to see the other members of the Physiological Department. The mechanician, a Mr. McCall, was building a treadle ergograph for Cathcart to do research in work physiology. Sitting on a bicycle seat, and quite erect, the subject was to work straight levers with the feet in the position usually occupied by bicycle treadles. The foot rest, on the short arm of the lever which was a 2 to 1 ratio. The lever can be lifted high or low and turns a wheel at the fulcrum which registers the total amount of lift through which the weight on the other end of the lever has been passed. The subject has also to do the negative work of letting the weight down according to the present arrangement. I was not especially impressed with the apparatus and took no photograph of it.

They had quite an intricate, but I imagine practical, reconstructed kymograph. The original part of it was made by Ednie, who was the mechanician for Schafer at one time and made much apparatus for other laboratories in England and abroad. Mr. McCall has arranged an electrically recording blood pressure apparatus which is worthy of note. A resistance wire is placed in one side of a glass U-tube about 1 foot long in an upright position. The rise and fall of the liquid in the manometer changes the effective length of the resistance wire and hence changes the current in the armature of a special galvanometer. This galvanometer carries a long pointer which writes directly on the kymograph record. The principle is useful and I think could be employed in making a respiration recorder for use while on the treadmill or when doing mild muscular work. (I at one time made preliminary experiments with a column of mecury using much the same idea but it did not turn out very successfully.) Mr. McCall had made his own signal magnets making an armature somewhat like that of the piano magnet. He had imagined that the principle of armature action was new and had thought of applying for patent.

Cathcart had a great deal to say about the publications of the Nutrition Laboratory. It was his opinion that there is no better work done, that we amass a wonderful amount of data, but do not know how to dress it up and put it out before the scientific world. He complained that everything is put down as of equal importance, there are no italics, and no index or summary. He said also that the reports were so huge that no one or very few read them, also that no one ever got into a fight with our statements because they were not pronounced enough. He argued stoutly for the position that a man should figure out his own experiments rather than turning them over to assistants and said that he himself could only work under the condition that he figured out one experiment before he did the next experiment. He felt that the Laboratory was not getting its due in present appreciation of its scientific work. After he had said many such things in a kindly spirit he stated himself to feel better having unloaded his burden of criticism. I was impressed that Cathcart retains a lively interest in the Nutrition Laboratory and has sincere wishes for its best success.

University of Glasgow, Psychological Laboratory,
Dr. H. J. Watt.

Previous to 1920 the Psychological Department has been housed in rooms loaned by the Department of Physiology. At present the Physiological Department being so overcrowded with students requires all its room. An old private mansion quite near the university buildings has been purchased and Watt is converting this into a psychological laboratory. He has just begun in his new quarters and hardly any alterations have been made. His apparatus equipment is at present quite meager. He thinks the prospects for the department are now good. Watt who is a British subject, was a prisoner of war in Germany for about two years. It seems that he was educated in Germany and he knew Dr. Marbe very well. He spoke considerably of this man and how that he by numerous conferences really writes papers of his students, not allowing his students to put forth any opinion but what represents his own views as professor in the department. Considerable was also said about the old laboratory of Wundt. Watt was at that laboratory for some time and volunteered as a subject for experimental work. The number of subjects who were available was very limited. In spite of this, however, he was not accepted as a subject. I understood him to mean that they were very choice in the matter of whom they used as a subject and did not want to get in anyone who might have introspections at variance with the views of the professor. He thought it was absolutely suicidal for an outside student to contradict or debate the point of view of the man in charge in a German laboratory, that is, as a usual thing. He found Professor Külpe was of a very different kind and gave his students much freedom. Watt did most of his work with Külpe and just before the war had returned to Germany to renew acquaintances there. Hence, with the declaration of war he was caught in Germany and imprisoned. He was kept for two years and had a very trying experience both mentally and physically. Although a large, well-built man, he has not entirely recuperated his physical strength. I was a little surprised when Cathcart was speaking with him that Cathcart should ask how he was feeling and if he was making any improvement, since Watt looked to be one who would have good health. Watt's wife came from England and lived in Groningen with Professor Hamburger. The latter was most kind and untiring in his efforts to secure the release of Watt. He wrote many letters to Rubner and other physiologists as well as to psychologists in Germany. Even Dr. Marbe tried to get Watt's release. Finally, Hamburger succeeded in arranging an exchange of Watt who was considered not fit for military duty with such a German prisoner interned in England. The German prisoner was sent from England but the German government delayed the release of Watt for several months. Finally he was given leave to go but had to be away within two hours of the notice. He had to secure a permit and some thing in the nature of a passport privelege. It almost seemed that it was a trick and the authorities thought he could not secure this and get away in the two hours' time, failing which they should have an excuse to keep him a longer while. His experience is very remarkable. Naturally he has not had time to do any research work since being home nor has he had sufficient energy.

Watt is particularly interested in the psychology of sound and has had some controversy over certain points with Titchener of Cornell University. He was much interested in learning more about Titchener as a man, his outside interest in music, etc., etc. Watt gave me copies of his most recent papers, one the article on psychology which was prepared for the Hastings Encyclopeedia of Religion and Ethics.

University of Cambridge, Physiological Laboratory.
Doctors Barcroft, Adrian, and Hartree.

Dr. Barcroft has recently built a large chamber with glass walls. (See figures 25 and 26.) It is divided into three main compartments arranged as the sketch below (See figure 27).

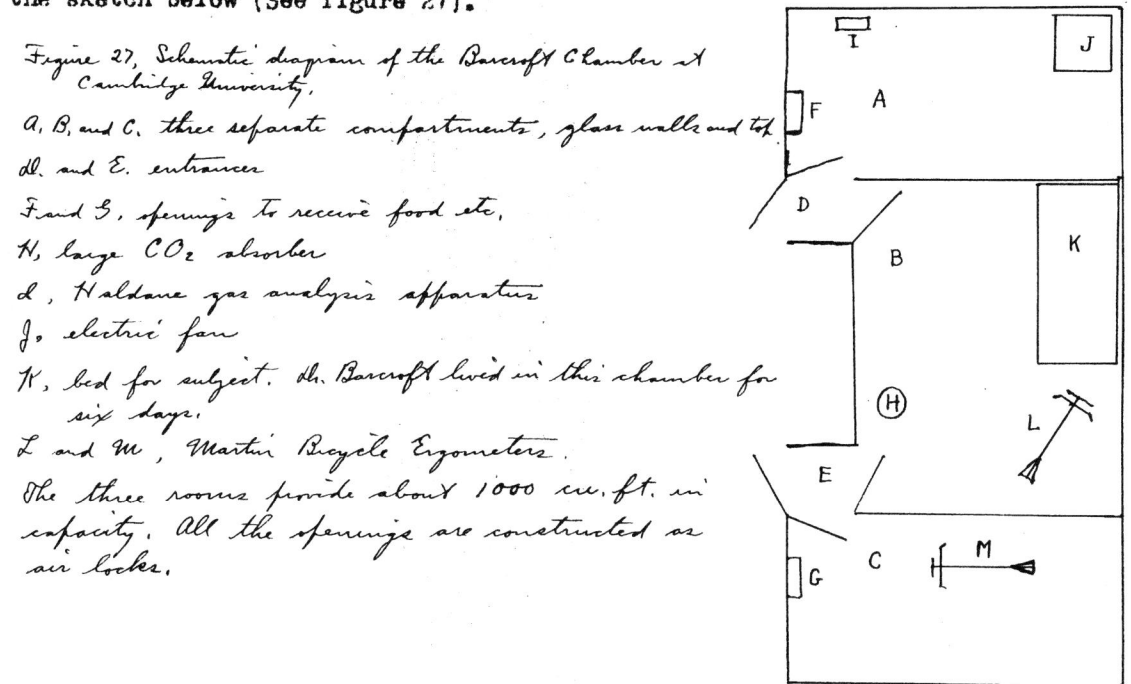

Figure 27, Schematic diagram of the Barcroft Chamber at Cambridge University.
A, B, and C, three separate compartments, glass walls and top.
D and E, entrances
F and G, openings to receive food etc.
H, large CO_2 absorber
I, Haldane gas analysis apparatus
J, electric fan
K, bed for subject. Dr. Barcroft lived in this chamber for six days.
L and M, Martin Bicycle Ergometers.
The three rooms provide about 1000 cu. ft. in capacity. All the openings are constructed as air locks.

This chamber is suitable for research on oxygen treatment using more than normal percentage of oxygen in the air and also for experiments with low oxygen. Barcroft has recently performed an experiment of the latter type living in the chamber continuously day and night for six days. He did not report to me all the details of the experiment. He had a big absorber bottle in the chamber and an electric fan to keep the air moving. Food was admitted and excrements send away from suitable openings in the base of the chamber. One such small opening can be seen in the lower left hand corner of figure 25. The bicycle ergometer was used for exercise and to measure the effects and efficiency for work under the conditions of decreased oxygen. In general the experiment was to ascertain if the lungs actually secreted oxygen according to the theory of Haldane and Bohr. Barcroft and Haldane do not agree in their interpretation of such results.

In the six days' experiment inside the chamber they made some psychological experiments on Barcroft with the assistance of Dr. Bartlett. These related mostly to the efficiency of memory for recall. Barcroft felt quite miserable the last twenty-four hours of the experiment.

I believe it was immediately following the end of the experiments living in the chamber that Barcroft made an experiment in the open room riding on the bicycle ergometer to determine the difference in the efficiency. In this latter part of the experiment he was breathing outdoor air and exhaling through a mask. He arranged very ingeniously an old hat with the crown mostly cut away for holding in place the accordion tubing from the mask. He referred to this arrangement as the "halo".

Barcroft, like many other of the experimenters in England, was using the Martin Boyle Ergometer for muscular work. This ergometer is made by C. J. Martin of the Lister Institute of Preventative Medicine, Chelsea Embankment, London. It has been described in the Proceedings of the Physical Society, volume   year . The rear wheel is fitted with an iron rim, rather heavy and having a flat flange of about two inched width. A canvas belt extends around this wheel. From one end of it a cord, passing around pulleys, is taken to the hook bflanspring balance mounted upon a board at the top of the upright in front of the subject. A cord from a second string balance mounted by the first one leads over suitable pulley to the other end of the belt. When the rear wheel is turned by the peddles, the scales drawn on by the tension of the belt which is in sliding contact with the rim of the rear wheel, show a difference in their scale of day, five pounds. The diameter of the wheel, which I believe is five and a half feet multiplied by the difference in the readings of the two scales balances in pounds and by the number of the revolutions of the wheel per minute, gives as I remember, the foot pounds of

work being done by the subject. I have not read the original description of
the apparatus nor do I know its constancy. I think, however, that the water
content in the belt must vary from time to time and even during an experiment
as prolonged turning of the wheel must cause the belt to get warm and dry out.
How much difference this makes in the amount of work figured according to
their formula, I do not know. It must make some but the apparatus has a great
advantage in its being very easy to arrange and to use. It seems always to be
in order and ready at hand. The subject can experiment on himself at least to
the extent that he syncronises his own revolutions with a metronome the rate
of which he knows, and he can see and read the scales as he is in position
riding the ergometer.

One subject on which I was interested to get information from Barcroft was
that of stop-cock grease for use with the gas analysis apparatus. Dr. Higgins
when returning from his visit to English Universities in 1913 brought back with
him a can of Jackson's stop-cock grease which proved very satisfactory for use
with gas analysis apparatus in this Laboratory. This was supposed to be available
through Baird and Tatlock. Barcroft had some of this grease which had
been given him by Mr. Allen of Baird and Tatlock Company but Mr. Allen would
not sell any of the grease. Barcroft did not know the reason. He informed me
that Mrd. T. N. Sturgess of the Clinical Department of the University of Bristol
had a formula and made up such grease that was very satisfactory. He did not
like to ask him what the formula contained but suggested that we buy some of
him and pay him for the formula. Barcroft thinks the stop-cock grease to be
made of rubber, vaseline, bees-wax, and paraffin, but is of the opinion that
the merits of the grease consist more in the way it is put together than in
the actual substances it contains.

Dr. Adrian.

Dr. Adrian's laboratory is in the basement and occupies one large room
with a dark room and one or two small adjoining rooms. This part of the
laboratory was planned by the late Keith Lucas and most of the instruments

Fig. 28 - Dr. E. D. Adrian in his laboratory preparing a demonstration for the Physiological Society Meeting.

Fig. 29 - Various pieces of apparatus devised and used by the late Dr. Keith Lucas and now in the laboratory of Dr. Adrian.

Fig. 30 - The capillary electrometer and signal galvanometer devised by Dr. Lucas. Note the stone pillars and slabs on which the apparatus is supported.

Fig. 31 - The light from the electrometer passes through a small opening (X) to the camera which is in the dark room

there at present which have been made and used by him in his remarkable researches in nerve and muscle physiology. Dr. Adrian, (See fig. 28), who had previously worked with Dr. Lucas, is continuing this field of research very worthily. The physiological building of Cambridge University is rather new and very adequate for this department. The building was given by a Merchant's Guild of London known as the "Worshipful Company of Drapers." Lucas chose the basement for his work and he wanted to be free from vibration in researches employing capillary electrometer. I was greatly interested to examine the Lucas apparatus with which I had been acquainted only through literature. One of his instruments, the pendulum contact breaker,(See A in fig 28)I duplicated with certain modifications in our own Laboratory. The original Lucas contact breaker of the pendular type is the large apparatus (B of Fig. 28). Along thr

Along the sides of the room heavy concrete slabs are arranged which make very splendid tables free from vibration.(See fig. 29). The room has a cement floor and in the center of the room or wherever was desired Adrian had groups of stone pillars which could be moved.(See figs. 30 and 31). They seemed to be large tile filled with sand and on top of these a very convenient arrangement of thin stone slabs which could be built up to any desired height for experimental purposes. The light passing from the capillary electrometer passes through a small slit in the wall, (X in fig. 31(, to the camera which was in t the dark room. Here Adrian could work with an exposed plate on a pendulum or other camera device for giving it suitable motion. The camera which they like best is the one shown in fig. 4 of Lucas' paper in Jour. 3 Physiol. Vol 39, 1909, p. 215.

One quite special piece of apparatus in the laboratory had not been described by Lucas.(See C. in fig 29 and also fig 32). It was a contact breaking device for giving successive makes or breaks as might be desired by the experimenter. The keys were two tuned steel strings on which one contact point was mounted.(See A and B. of fig 32). This contact point was moved outward into contact with the other point and as soon as released by the cam breaks

GREAT BRITAIN

Dr.
Noguès (M.D.)

Bulla stereoptic camera Zeiss Tessar 6.3.
6×13 cm. L. Gaumont Paris
                Stereo-Spido.
Fournier & opposite
     Brd Beaumarchais
          hop. 11^e

Magnéta Baillie Freres. Magnetidal?
28 Boul de Villiers
       Levallois-Perret.

~~Langlois~~
  Langlois

. School Medicine.

Platinum wire .4 mm. about 10 turns
10 volts.
The little electric furnace

Dr. Bulla Marker telling me
about Sound Photography.
June 2 - 1920

10 cm his
50 cm distance
AC 1 K.M.

Bull
11 Brd Delessert
Metro-station Passy.
4th floor
3rd button in lift

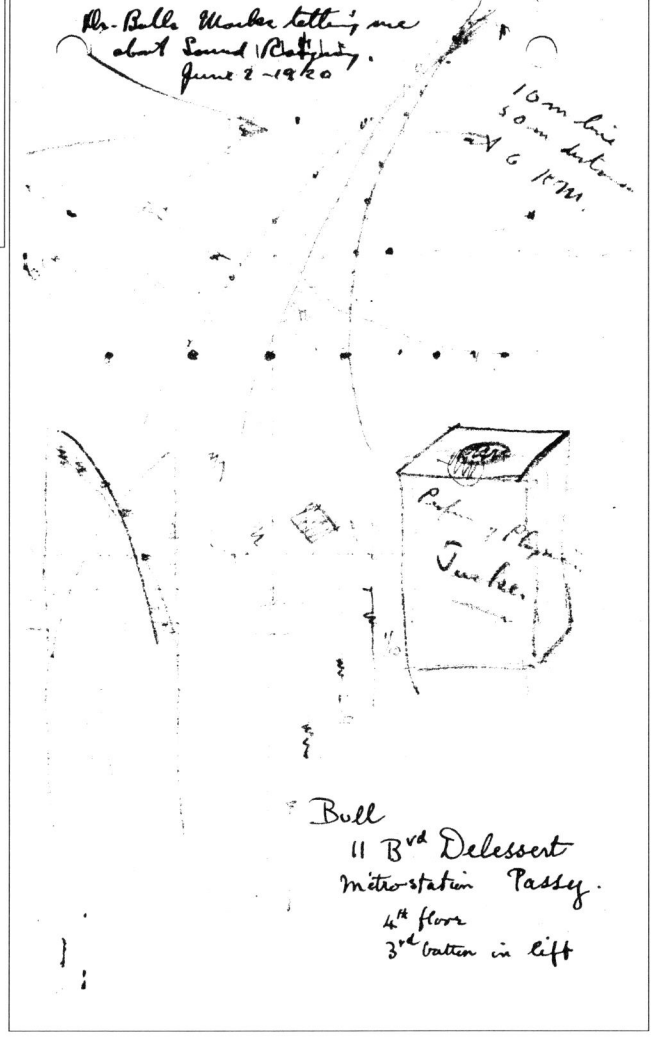

contact by its own tension of the string. A series of wheels mounted of the right hand side of the device (C, in fig. 32) allowed one to use all of the makes ar breaksoeresserysedcond one or every third one or one for every revolution of these accessory wheels. Adrian asked me for the film of this picture (fig. 32) for this apparatus and said that he expected to describe it before long and would like to use the photograph.

Adrian not only has grear ability as a man in research but he is gifted in administrative affairs and very prominent in his own college(Trinity College) and also in the general work of the University serving on a large number of boards. During the days that I sppnt so pleasantly as his guest at Trinity College, I was impressed that from one quarter to a third of his time must have been taken up by these various board meetings. In fact, Adrian told me that he had practically no opportunity for research except during vacation periods. Very recently a new Board of Psychological Studies had been organized and he was serving on that.

While in Adrian's laboratory taking photographs I had the pleasant experience of meeting Professor Sedgwick of Massachusetts Institute of Technology, andhangbangefessor in England, who happened to be that day at the physiological building an accompanied by an official of the University, looked in for five minutes on Adrian.

Dr. Hartree.

Dr. Hartree is physicist and came to the Department of Physiology in Cambridge to help Dr. A. V. Hill who, for a number of years has been doing most important work in nerve physiology, some of it in collaberation with Lucas and Adrian. Hill has very recently gone to Manchester as professor of Physiology and Hartree is continuing the work in Cambridge under his direction. He is continuing the research on the heat production of stimulated nerve trunks and they have the most complicated set-up of small electrical apparatus with which I came into contact anywhere. Hartree showed me the construction of the thermo piles which they use in their nerve work. The nerve is laid across the thermopile when stimulated and when measuring it for heat productions, due to the passage of conducted disturbances. These thermopiles which have been described in some of Hill's articles are extremely delicate things to make requiring great care in the cutting of the lengths of the wire and in soldering the two elements together. They embody many turns of wire and I should think a hundred or two soldered joints. Hartree described to me the new type of Paschen galvanometer which he was using. It is a quartz suspension which is attached at the upper end. At the middle of this fiber is a ~~small fiber is a~~ small mirror which is stuck on and at the end a very small damping vane.

Schematic diagram of the Paschen Galvanometer
A, point of attachment
B, quartz thread
C, and E, groups of very tiny bar magnets
D, small mirror
F, damping vane.

At a point half way between the suspension and the mirror and again half way between the mirror and the damping vane thirteen very tiny magnetized bits of iron rod are stuck on to the glass fiber. A very slight amount of current is the field coils causes the glass fiber to turn. Hartree had experienced considerable difficulty with current leaks into his galvanometer arrangement. We discussed the proper mounting of electrical instruments on dry wooden or porcelain bases and experimented some with his apparatus. It seemed impossible to find the source of his trouble.

University of Cambridge, Psychological Laboratory.
Doctors Bartlett and Muscio.

I had learned from Sir Walter Fletcher at the meeting of the Royal Society that Myers had left the Psychological Department of Cambridge and taken up his residence in London where he was practising psychiatry, (a line of work with which he had much success during the war with shock cases) and laboring for the establishment of a National Institute for Applied Psychology and Physiology in England. It seems probable that Dr. and Mrs. Myers preferred to live in a large city rather than in Cambridge and this is one of the reasons why he has left the teaching of psychology.

The Psychological Laboratory is one wing of the new Physiological Laboratory. This wing was separately financed. Myers secured by private subscriptions the money for its building. It was planned by him and is very excellent in its arrangement for instruction and research work. In the matter of apparatus they have not as large an amount of stock pieces as I found in the University at Edinburgh. The apparatus that I was especially interested in looking at was the Kraepelin ergometer with a shot binding pawl arrangement to obviate the negative work, the Rivers-McDougall dotting machine, which by the way, was quite difficult to get into operation, and an outfit for psycho-galvanic reflex work. They have quite a satisfactory sound-proof room and considerable apparatus, (home made) for working with binaural sound localization. In a room devoted to sound they have an immense collection of phonograph records of primitive music secured from a very wide range of tribes of people and much of it I believe by Rivers and also Myers when on anthropological expeditions. This material remains to be worked up from the standpoint of the psychology of primitive music. In fact Myers had on hand a great amount of incomplete research work some of it springing out of problems of the war. It seems that if he leaves this in its present state of incompleteness and it cannot be utilized by Bartlett or students in the department, it will be a heavy loss.

B Bartlett is himself working at present with non-experimental material. He has an anthropological problem involving the accuracy of report and quotations He had many publications about and was tracing the history of certain legends and stories as handed from one tribe and people to another noting the changes which were made as the story was passed on. Bartlett is a large, quiet man and now in charge of the department. He showed much interest in American psychology.

Dr. Muscio.

For the last year or two Muscio has been working under the direction of the British Industrial Fatigue Research Board with headquarters at 6 John Street, Adelphi, near Charring Cross, London. The secretary of this board from whom information can be secured about its activity is Dr. D. R. Wilson, Sherrington is a member of the board and Lee of Columbia University expects to spend the year of 1920-21 in connection with this work in England. Muscio described certain research work that he had done in one or two factories in connection with the relative fatigue developed in the forenoon and afternoon's work as compared with the forenoon and afternoon preformance when the subjects were not working and hence not fatigued. He showed that the difference between the forenoon and the afternoon was about the same in both cases and a larger increment that the difference between the level of the two curves for the working day and for the day on vacation. The test employed was that of the "Match Board" described in the book by Kimball entitled "Chosing Employees by Test". The "Match Board" requires quick coordination in placing small pegs in small holes in a regular pattern. Both Muscio and Bartlett seem to think it a good test to use. Muscio at one time when he was going from Australia to London to work with the Fatigue Board visited the Nutrition Laboratory and saw the pursuit pendulum, the pursuit meter and other apparatus. He later wrote for a description of the pursuit pendulum and signified his intention to make one. I sent him photographs and data but found he had not up to the present, built such an apparatus. He urged me to visit the officers of the Fatigue Board in London where I would be able to find more concerning Myers plans for the National Institute for Applied Physiology and Psychology. I was unable to make this visit, and up to the present I have received no publications from this board.

At the psychological laboratory in Cambridge they had a way of keeping all the doors locked so that unless you had the personal guidance of Bartlett or Muscio it was impossible to go around and further examine pieces of apparatus that I should have liked to have seen in more detail.

University of Cambridge, Biochemical Laboratory,
Doctors Hopkins, Peters, and Grey.

My conferences with Dr. Hopkins were limited to meeting him in a social way at a Trinity College banquet and sitting by him at the meeting of the Physiological Society. However, under the guidance of Peters, I had an opportunity to visit that Laboratory which is an old building remodeled for the purposes of their department but they must soon have better quarters for this work. I saw the extensive rat colony on which Hopkins has done and is doing important work with the problem of vitamines. I met the assistant in charge of this work, Miss Stevenson, a clever woman.

Dr. Peters' research work was then with the organism, Paramecium. He had by great patience and perseverance developed a pure strain of Paramecium all from one organism and had hundreds of cultures of this strain which he was subjecting to various types of environmental treatments. We mentioned that it was very easy to kill them if the water was not pure and discussed the power of adaption in the organism.

Dr. Grey impressed me as a most excellent workman in biochemistry. He has received training from the Lister Institute and also from the Pasteur Institute. He had many methods and short cuts which seemed to me probably quite valuable for workers in this field but I could not well understand or remember them. For example, he had much to say about the use of silica flasks and has invented a special cork for such flasks. This cork allowed evaporation out of the flask and was something like the diagram below.

Diagram of Special Glass Cork
a, neck of flask
B, hollow glass cork filled with loose asbestos.
C, hole in cork to permit evaporation.

It to him had a great advantage in that it did not let any loose bits of asbestos fall down into the flask and into the material that he had there. Another thing was the impregnation of asbestos with palladium. Such a little wad of impregnated asbestos was connected to the electric current, placed down in a flask, caused to glow and dry out the flask making it "anaerobic" according to the standards of Laidlaw. Grey had also a routine for making up certain standardized solutions in quantity sufficient for a year's use which he thought of great impostance for economy in the laboratory. He said that the wooden floors of this old building were so impregnated with chamicals and gas of one character and another that he was limited in not being able to work on certain problems in that laboratory. I regretted afterwards that my visit with Grey was so short and my understanding so limited.

Meeting pf the British Physiological Society.

(Saturday, May 15).

The program of the meeting of the Physiological Society is attached herewith. I shall not attempt to give a summary of the various demonstrations and papers which were presented. Under the topic,"Two measures for muscle coordination" I described the pursuit pendulum and the test for static control both of which I had developed at the Nutrition Laboratory. By the use of lantern slides I demonstrated the apparatus and showed some illustrative data. The Society received me very kindly. I was interrupted during the presentation by friendly questions from both Halliburton and Bayliss.

The meeting gave me an excellent opportunity to see many of the British physiologists together also to meet some whom I had not been previously able to visit, notably among these A. V. Hill and W. H. N. Rivers. With the latter I discussed alcohol experimentation and he kindly invited me to come to Cambridge and spend a day or more with him at St. John's College in August. His invitation I was unable to accept.

Following the meeting of this Society the return trip to London was in the nature of a little scientific meeting in itself since Dr. Haycraft (J. B.), Professor Dale (H. H.), Dr. Evens (a member of Dale's department), Dr. Cullis (a very capable woman) Professor of Physiology in the Woman's Medical College of London, who had previously worked with Dr. Brodie, Dr. Mallenby, and myself were in a compartment during this trip. The discussion centered largely on the topic of psychical research and I was much impressed to see how interested this group of scientific people were in that problem and there was also some discussion of the effects of alcohol. Dale made a significant remark, I thought. He said in reference to alcohol experiments it is foolish to ask,"Did they disguise the taste ?" because it is impossible to disguise an alcohol dose if given even in moderate quantity, particularly after the subject has taken it more than once.

It seems to me probable that British Physiology owes a great deal to the stimulus and impetus which comes through the monthly gatherings of the Physiological Society. While the meetings, of course, have a social side yet they come so relatively frequent that the social side is not the most prominent. There is no other meeting of an allied society or branch of science at the same time. Their scientific papers and discussions therefore assume more importance, in fact, the chief place, and I think make for real scientific progress.

Cambridge and Paul Instrument Company.
Dr. Whipple.

Dr. Whipple was particularly interested in the work on skin temperature which has been going on in this Laboratory. They have arranged an apparatus for measuring the skin temperature continuously using what is known as their "thread recorder". A thermocouple for taking the skin temperature is connected to a galvanometer which carries a long arm. Below this arm is a paper for recording the deflections of the arm. Between the arm and the paper is a thread much like a typewriter ribbon. A clock operates a tripping device and causes the pointer to be quickly depressed against the thread and upon the paper record and so leaves a dot. The thread is moved along much as a typewriter ribbon so that the same place is not continually struck. It is an interesting arrangement for recording purposes.

Under Whipple's guidance I saw their latest improvements on their string galvanometer. They have changed the type of mounting for the telescopes,(See fig. 33, D and E.) making them much more rigid and more easily adjustable than the type of Cambridge string galvanometer for example which we have here in this Laboratory. They reported that the point source electric lamps were exceedingly practical for use in protection work with the galvanometer outfit and gave satisfactory definition. They are manufacturing a new and simple paper camera which has grown out of their product made during the war. It was one feature of the sound ranging apparatus later to be described which was made by Dr. Bull of the Marey Institute of Paris. In this new form of camera they run the motor continuously but start the paper by tightening the belt. This did not seem particularly satisfactory since the belt must slip in the pulley in between periods of feeding the paper and is apt to cause the pulley and paper to crawl slowly. They were experimenting on different forms of devices for starting and stopping the paper feed and asked me to send them pictures of the modifications which I have made on their paper camera in this Laboratory.

Whipple urged me to visit Dr. Woodhead in Cambridge informing me that he was

92

greatly interested in the measurements of skin temperature and also in any alcohol research. There was not time for such a conference, much to my regret.

University of Oxford, Physiological Laboratory.
Professor Sherrington, Dr. Bazett and Dr. Douglas.

In Professor Sherrington's department they had the same problem as found in the other universities in England, that is, overcrowding the students at the present time with consequent interruption of research. "A photo of Sherrington at his desk is shown in fig. 34. Although the laboratory building is old, it has been modified in certain particulars and is still quite adequate to the teaching and research needs. The room in which the practical class works, see figure 35, is lighted by large windows on two sides and also by a skylight which is really a continuation of the windows on one side of the room. There are six or seven large long paper kymographs so that twelve or fourteen advance students can be accommodated at the same time and the working conditions are especially excellent.

Sherrington had quite a lot of new apparatus in process of construction and employs a very good mechanic. He had recently built an elaborate arrangement for successive electrical stimulae. The contact points were mounted on tuned steel strings and arranged so that by electromagnets they could be made to vibrate at the desired rate. The apparatus also employed a number of specially constructed electromagnets which operated armatures somewhat like the piano magnets which we have at the Nutrition Laboratory.

There was in this laboratory the largest myograph which I have ever seen. It was mounted at one end of the room behind a large glass window which could be raised when the myograph was to be used. As a myograph it leaves very little to be desired but Sherrington does not like to have his records described on the arc of a circle. It makes the careful measurements of curves difficult in reference to the time line which accompanies such records. He was engaged in devising apparatus which would give him the movement of a kymo-

Fig 34- Professor C. S. Sherrington at his office desk, Physiological Laboratory, University of Oxford. He was working on some drawings of the semicircular canals.

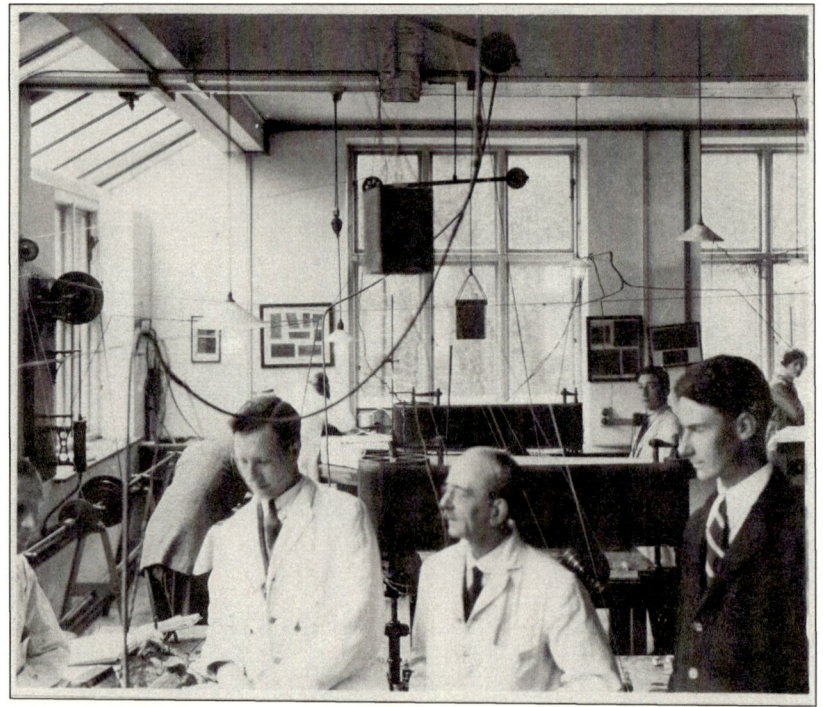

Fig. 35- Professor Sherrington's laboratory for the course in advanced physiology.

Fig. 37 – Dr. H.C. Bazett (right) discussing his course in Clinical Physiology with W.R.M.

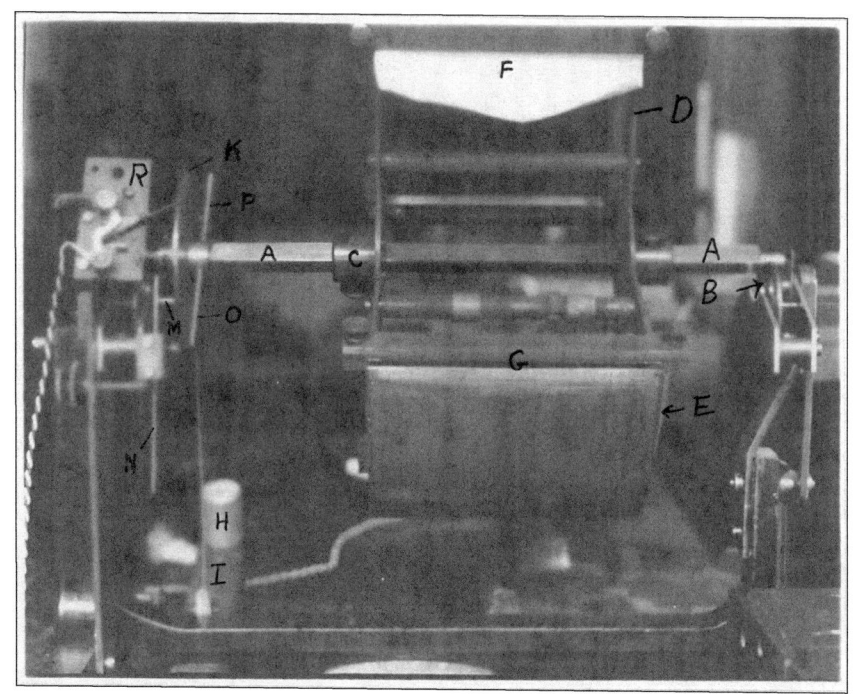

Fig. 36 – A new form of myograph devised by Professor Sherrington to take the place of the usual pendular type.

graph paper at fairly high speed and in such a way that the record inscribed along straight lines. (See fig. 36). On a shaft, A, mounted on Atwood bearings, B, he had a drum, C, of special design such that one half, D, balanced the other , E, and that short kymograph papers could be easily prepared and mounted on D and E. Two papers are used and each end of a paper is held easily in a clamp as shown at F and G in the photograph. The drum was caused to revolve by a weight, H, (he could, of course, use weights of different mass) operating upon a lever arm, D. The main shaft has a wheel , K, at the left and from this wheel two studs project. The lever arm on which the weight acts is fitted with a long hook, L, which takes hold of one of these studs on K. The other stud on K is seen in the figure at H; it is held by a pawl, M, in the foreground. When this pawl is released by hand, the weight acts upon the shaft giving the whole thing a turn over. When the weight has come to the lowest position in contact with the platform, the hook, L, releases the stud upon which it has been pulling and the drum goes around until the pawl catches that stud and locks the drum in position. The apparatus is so arranged and so light in its action that the period of acceleration of motion is very short. Following this period, there is, according to Sherrington, quite a long stretch of movement which is at a constant speed. Just how long this constant motion lasted, I cannot say. I think it is possible that the levers which are to write on the kymograph have to be so adjusted as to come into contact with it a short period after it starts to move. The shaft carries arm projections for opening a circuit, see O and P, and the key R in the figure.

Sherrington is a member of the advisory committee of the Central Control Board (of the liquor traffic in England). He is quite keen on the subject of alcohol research and questioned me concerning the work at the Nutrition Laboratory. He thought that a test of static control such as the standing test was a very good one to use in reference to alcohol investigations on human subjects. He made remarks about the changes in the gait of a man after being given alcohol, also concerning the influence upon vision and upon marksmanship. He suggested that I call at the offices of the Central Control Board (of the liquor traffic), 134 Piccadilly, London, and he gave me a note of introduction to the secretary, Dr. Sullivan. I found later that he had reported our conference to Lord D'Abernon for the latter wrote me of his conference with Sherrington and asked that I should visit him in London, to discuss the matter if possible. A conference was tentatively arranged but could not be carried out due to the fact that Lord D'Abernon was sent as British Ambassador to Germany, and, at the time I was in Berlin, had gone to Poland, in reference to political matters.

When I was with Sherrington, he said that he knew about my alcohol work and showed me the short paper, the preliminary report given to the National Academy. He made no reference at all to publication 266, although I am certain it was sent to him. It was interesting for me, although possibly just a chance occurrence, to note that he had the short paper on hand, had read it, and referred to certain points in it but the monograph, the longer article, seemed to have made no mark on his memory.

One cannot help but be impressed with the genuineness and kindly eagerness of Sherrington. He is the very anthesis of self assertion and so ready to point out things of interest in your work rather than discussing the importance of his own. His slight tendency to lisp and hesitate in speech only add to the interest with which one hangs on his words.

Dr. Bazett.

Dr. Bazett is giving a course in Clinical Physiology which is on the human side as nearly as possible a parallel at Sherrington's course in Mammalian Physiology. He uses or intends to use many measurements on human subjects not only involving strictly physiological functions but also some psychological functions and we found much of common interest for discussion. Bazett is a very earnest and hard working men (See fig. 37). Some of the measurements at present in use by him may be enumerated: (1) Vital capacity, measured according to the method of Dreyer. Dr. Cairns, an Australian, is working under Bazett's direction in research on the problem of measuring vital capacity. He finds, of course, that the vital capacity measured in liters bears a relation to the amount of CO_2 in the lungs. If the subject makes a deep inhalation preparatory to measuring his vital capacity and holds his breath for a time allowing the CO_2 to build up, the amount of air that he is able to expel will be larger and hence his vital capacity measure increased. A second point which Dr. Cairns is working out is the effect of resistance. He thinks that if there is a certain amount of resistance to exhalation, for example, exhaling through a small tube or a tube with cotton wadding in it, the vital capacity measure is larger. (2) Exercise power. They use two dumbbells of ten pounds weight each lifted from the floor to the highest convenient point above the head with arms extended, and measure the blood pressure and heart rate as quickly as possible after the exercise, noting the progressive changes following the exercise. (3) The Flack Blowing Test. The subject blows through one nostril, holding the other closed with his finger and endeavors to hold a pressure of forty millimeters of mercury for forty seconds. The experimenter watches the pulse rate change immediately after the test. Dr. Bazett remarked that they have not found much effect yet but think there us a little difference between the results obtained on Saturday and Monday. He was interested to learn the way in which we, at the Nutrition Laboratory, have commonly taken the electrocardiogram during exercise. (4) The Total Ventilation of the subject per minute for a three minute test. (5) The Standard Electrocardiogram. Bazett thinks that this may possibly be used for a measure for the dilation of the heart indicated dilation and hypertrophy that the heart has shown that is length between of Dr. Meakins. A subject of research

in which Bazett is especially interested is the time value of the waves of the of the electrocardiogram and the ratio of these values to the rate. He measures from the end of the T wave to the end of the next T wave for securing the hos cycle length and finds that from the start of the R wave to the end of the T wave equals .365 times the square root of the time from the end of the T wave to the end of the next T wave, i. e. of the total cycle length. He referred me to the work of Dr. P. H. Dawson, who found the square root function of the carotid pulse, that after a long systole, the diastole is often short, while after a long diastole, the systole is also long. (6) The use of the "Duboscq" Colorimeter. The above mentioned tests are selected with a view to measuring physical fitness of a patient.

A line of research in which Dr. Bazett was particularly interested had to do with the keeping for a long time of decapitated animals, for the study of a variety of physiological problems. He was using cats and found it possible to put the animal in a vessel which was floated in a water bath maintained at a relatively constant temperature. He used a rectal thermometer in connection with one of the Cambridge Instrument Companies thread-recording galvanometers, to obtain continuous records of body temperature.

During the war, Bazett did considerable work on problems with the personnel of aviation. He was particularly concerned with the cardio-vascular reactions of the pilots. He is well acquainted with Boston having worked a year or more with Dr. Porter at the Harvard Medical School. I found as many problems of common interest with Bazett as with any other one man whom I visited during this trip.

Dr. Douglas.

Dr. Douglas showed me the apparatus and methods which he had recently employed in his study entitled "Respiratory Exchange of Man During and After Muscular Exercise." In this he had used four "Douglas Bags" in succession mounted conveniently upon a rack and had employed the bicycle ergometer for the muscular exercise. I saw his large respiration chamber. (See figs. 40 and

Fig. 38 - Dr. C. G. Douglas in his laboratory at the Physiology Department, University of Oxford. He was looking up a point in Carnegie Inst. Pub. 187.

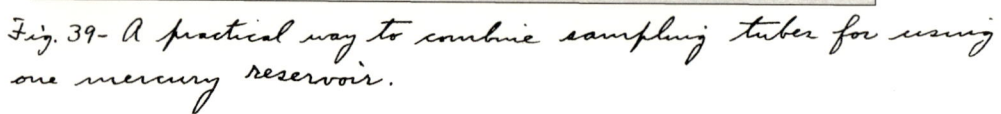

Fig. 39 - A practical way to combine sampling tubes for using one mercury reservoir.

Fig. 40 - Dr. Douglas sitting inside his large respiration chamber. A reflection in this picture gives the appearance of an outside window at the back. I do not recall one but there is such in the chamber in Haldane's Laboratory.

Fig. 41 - The CO_2 absorber and ventilating system of the respiration chamber used by Dr. Douglas.

41). Its relative size can be fairly accurately judged by comparison with Douglas sitting at the window inside the chamber. It is lead-lined and has two doors, one large and one small. The air enters at the bottom and is sucked out at the top passing through the large absorber shown in the figures. Aside from this chamber, I judged there was no new apparatus in Douglas's laboratory with which my colleagues had not already become acquainted.

Douglas recommended especially that the gas sampling tubes should be mounted on a rack and joined together according as in figure 39 using one outside bulb of mercury. He had found this arrangement much more economical of mercury and very practical when later making gas analysis.

Douglas called my attention to what he thought to be a misprint in the book of Benedict and Cathcart, Table 68, the last figure in column "Oxygen absorbed per min", 3265, which he thought should be 2265. While he was looking up this point I took the photograph of him. (See fig. 38).

Haldane's Private Laboratory, Oxford.

Dr. Haldane.

By building an addition to his beautiful residence on Linton Road, Haldane has provided himself with a private laboratory, very well suited to his research needs. This laboratory is reached by an outside entrance or through a short hallway from Haldane's office which latter is a large room in the residence building. The photograph of Haldane (fig. 42) shows him at his office desk and the door entering his laboratory is in the background. Some idea of the laboratory may be gained from figure 43. In one corner he has built a large respiration chamber which is practically a duplicate of the one seen in Douglas' laboratory. It has some minor improvements, is lead-lined and one of its windows look out on a pleasant garden. A rather large window opens into the laboratory. The door through which the subject enters can be opened also from the inside.

Two small rooms, (a shop and a store room) open off the main laboratory. There is a large amount of respiration apparatus, the only new thing, I understood to be, the blood-gas analysis apparatus, which had been arranged by Haldane, (the same apparatus that was being used by Dr. Meakins at the Royal Infirmary in Edinburgh). The mechanician was employed in making this apparatus for sale.

Haldane was being assisted in his researches by his son, Dr. Jack Haldane, and by Dr. Davis from Adelaide, Australia. The latter is going to work with Meakins the following year. They were engaged in research on the CO_2 production of the kidney. They used a blood-gas analysis apparatus also for work with urine examining the bi-carbonate in alcoline urine.

Haldane had recently made an extended experiment (somewhat like that of Barcroft), living for some days in the large chamber and decreasing the oxy

Fig. 42 – Dr. J. S. Haldane at his desk in his residence on Linton Road, Oxford.

Fig. 43 – The physiological laboratory in Dr. Haldane's residence. Dr. Haldane was assisted by his son, Dr. Jack Haldane, (center) and by Dr. H. W. Davis of Australia.

gen gradually, until it was equivalent to a barometric pressure of approximately 320 milimeters. They endeavored to hold the oxygen at this level and remarked that when the CO_2 had been pumped out, there were strong subjective symptons. The subject was unable to write and he was very stubborn and hard to deal with. He repeated his instructions to them many times, signifying that he was not capable of clearly seeing that they understood the instructions. Their research had to do with Haldane's theory of the secretion of oxygen by the lungs. For muscular work in this laboratory they also employ the Martin Bicycle Ergometer. Haldane, who had visited the Nutrition Laboratory on more than one occasion, exhibited interest in our work, much to our pleasure, and supplied me with a large number of reprints.

University of Oxford, Pathological Laboratory,
Professor Dreyer.

The work in progress in this laboratory seemed almost wholly of a physiological character. I remarked to Dreyer that it seemed to me he was running another department of physiology at Oxford University. In reply he said that Pathology must have certain standards from which to proceed and that there were many standards of normality which Physiology has not satisfactorily established. He, therefore, found it necessary for his own work to endeavor to establish these before proceeding to consider pathological phases of the questions. His present interest and research is especially in reference to certain physical measurements such as weight, sitting height, and vital capacity and the possibility of predicting metabolism from these.

He thinks that the accurate measurement of sitting height can best be obtained when the subject is seated on the floor or on a board with the knees drawn up. He recommended that the subject, bearing his weight on his knuckles, should lean forward and press his os sacrum as close as possible against the vertical scale at his back, then straighten up to full height. The rider on the centimeter scale was of special construction. The part carrying the pointer is so arranged that on no account would it press hard on the subject's head. This seemed to him important that the subject should not bump the top of his head against some solid or more or less immovable rider on the scale. His rider was of a form indicated by the accompanying sketch.

Diagram of Special Rider for Measuring Sitting Height.
A and B, measuring scale
C, rider for rough adjustment
D, rigid arm carried on C.
E, arm carried on light hinge, to come into contact with the subject's head.
F, point at which to read scale after placing slight pressure on E.

Fig. 44 — Professor Georges Dreyer, Department of Pathology, University of Oxford. (It was a dark day and we were upstairs by a skylight. A slide rule, "my constant companion" he said, was in his hand.)

Fig. 45 — Mr. G. F. Hanson, an American, working with his newly developed gas-mixing meter.

Fig. 46 — Mr. H. F. Pierce with the splendid low-pressure chamber installation designed and built by himself.

Fig. 47 — Mr. Pierce and Mr. Hanson inside the chamber, indicating the commodious space for experimentation.

The experimenter, with his finger, places a little pressure the lower part, B, after the general position has been adjusted, and so obviates any change in reading due to the stiffness or piling up of the hair and also does not clamp down solidly on the head. Dreyer and I took several successive measurements using each other as subject and found these to be remarkably constant, varying only one or two mm. in a total weight of 92 cm. He thought the sitting height, taken in this way, is probably the most constant and repeatable physiological measure. It, as well as the measurement of weight, is usually taken without clothing.

The chest circumference measure is taken by tape placed directly on the skin around the chest just on the nipple line, in the case of male subjects. While the subject is being measured, he stands with his arms hanging loosely down at his sides, breathing normally, and is encouraged to talk as the muscles are then more satisfactorily relaxed. The standing height, without shoes, is usually recorded also on their blank, a sample of which is attached on the next page.

Dreyer showed me a paper which he had just written in collaboration with Dr. Burrell on the subject of "Vital Capacity Constants Applied to Pulmonary Tuberculosis". This paper (See Lancet, June 5th, 1920) based on the data from 150 or more cases, shows quite conclusively that Dreyer, with only vital capacity measurements, and not having seen the subjects, was better able to tell whether the subjects were improving or declining than was Burrell, who was in personal contact with the patients.

I was shown their method of measuring the vital capacity by a student, Mr. Hobson, who is working on the collection of this data under the direction of Dreyer. The subject does no forced breathing prior to the measurement. He stands, is told to inhale as much as possible and then to exhale rather slowly. He does not see the result. The experimenter reads the meter, records the amount, and sets the meter to zero. The subject tries again, and makes five successive trials, with about one-half to one minute in between each. Usually,

107

(Area Intentionally Blank)

Further descriptions of method and data are given in Dreyer's papers particularly in the little book just published "The Assessment of Physical Fitness: By Correlation of Vital Capacity and Certain Measurements of the Body", with Tables. by George Dreyer and George F. Hanson , 1920.

the last blow is higher than the first and the third or fourth will probably be the best. They take as their measure the maximum found in the five trials and have discovered that in a large number of observations, the highest reading of the five was about five per cent greater than the average of the five observations. They use a Verdin dry meter made by G. Boulitte, Paris, because of its low resistance and portability.

No one whom I met was more lavish in his praise of the Nutrition Laboratory and its work than was Dreyer. He had lately been making use of a large amount of our metabolism data in working out a prediction formula for metabolism. He has published on this topic in Lancet, August 7th, 1920. I regret that my picture (See fig. 44) of him does not show the slide-rule which he held in his hands and which he said was his "constant companion". The day was dark and for the picture, he left his desk and went upstairs, standing near a skylight.

Mr. George F. Hanson, a very clever American, was working in Dreyer's department and had developed and patented a meter for accurately measuring the mixing of two gases. The principle of this meter is explained in the patent specification. It seems that it will probably be useful in the mixing of gases for research in respiration physiology, also under conditions where the gases are used for anesthesia and for the mixture of gases for use in gas engines. A picture of Hanson and his apparatus is shown in figure 45.

Dreyer and some of those at present working with him were employed in research in connection with problems of aviation during the war. Such was the case with Mr. Pierce who had largely to do with the design and building of the low pressure chamber at Mineola, Long Island, New York. He had previously worked in Cannon's department at Harvard. A similar chamber for use in testing pilots in their ability to withstand high altitude effects was built in France. It was practically a duplicate of the Mineola Chamber, being a cylindrical steel tank about eight feet in diameter and ten feet high, standing on one end. (See fig. 46). It is entered through a full sized door in the side and forms quite a convenient room, large enough for four or five adults to be seated and at work. (See fig. 47). The reduction of pressure is brought about by a motor driven vacuum pump of about ten horse power capable of rarifying the atmosphere within the chamber to a barometric pressure 140 milimeters of mercury. The inside of the chamber is finished in flat white and is very conveniently light. There are several windows of thick glass through the walls of the chamber. One round window is in a convenient position for the man who is operating the motor driven pump and controlling the barometric pressure inside the chamber. (See figure 46). At the close of the war or soon after the armistice, Pierce succeeded in having this chamber moved from France to the Department of Pathology at Oxford to be used for research purposes. He has shown great ingenuity in arranging the equipment of that place. He was using goats as subjects in the chamber and was studying the reaction of the blood to altitude changes, noting the progress of adaption to high altitudes. He had a very simple and he thought extremely accurate method for drawing blood samples and making blood counts. Pierce informed me that he expected to spend the summer at the Research Laboratory of the General Electric Company, Schenectady, New York and upon returning to Oxford in the Fall, would utilize the latest advances in certain lines of science and build a respiration chamber for the Pathological Laboratory. The young men working with Dreyer were, I think, about the most enthusuastic research workers that I met. This judgment may be influenced by the fact that they were Americans. The output of Dreyer's department will, I think, be of much interest to workers at Nutrition Laboratory.

110

University of Oxford, Psychological Department.

Dr. McDougall.

Dr. McDougall did not show me any laboratory or any apparatus. I understand that previous to the present crowding by medical students McDougall had a room or rooms in the Department of Physiology. He had recently been spending his time in Zurich, Switzerland with Dr. Jung studying his method pf Psycho-analysis, a line of work which McDougall pursued during the war. He was at present giving a course of lectures at Maudsley Hospital in South London where they have recently established a Psychological School of Medicine. The place was opened in 1914 with a sum left for that purpose but was used as a military hospital during the war.

In reference to the alcohol research and the effects of alcohol, McDougall felt that a moderate amount of alcohol very frequently relieved nervous tension and enabled an individual to proceed with his work with much greater comfort even though he did not do _quite so much_ work. He thought this an important phase of the alcohol effect which had been largely neglected, by those over-enthusiastic for the prohibition of alcohol. He said that he expected to write up this matter as a member of the advisory committee of the Central Control Board (of the liquor traffic in England). McDougall, anticipating taking up residence in Cambridge, Mass. at the beginning of September as head of the Department of Psychology of Harvard, asked me a number of questions as to living conditions in this part of the world. We are delighted to have him come to the United States.

France

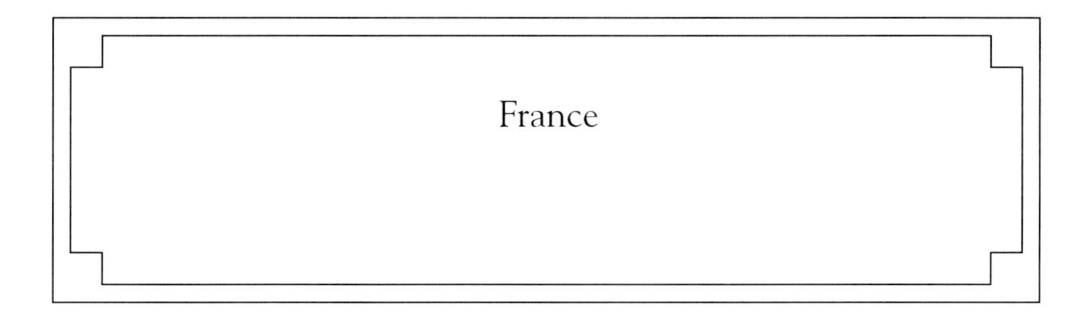

Marey Institute.
Drs. Bull and Nogues.

My first question after meeting Dr. Bull was to ask him about the report he made to the Societe Scientifique d'Hygiene Alimentaire and of their progress in building a calorimeter. I was astonished to learn that Bull, although returning to France approximately more than six months before the beginning of the war, was never asked to make a report. I understood him to say he was invited to attend one of their board or council meetings following his return, that he went with all his material both pictures, diagrams and notes, but was not called on or given an opportunity to make a report and since had never heard from them.

During the war, Bull's scientific ability was used in developing the sound-ranging apparatus employed by the British on their front. The problem was suggested to him by a man who visited the laboratory with a scheme for using several men in listening posts as reaction agents, each one pressing a separate reaction key when he heard a certain enemy battery. This man thought that with trained observers for reacting the time difference between the reaction signals marked on a kymograph would allow them to calculate the position of the enemy gun. It was only natural that Bull should think of a method of using the string galvanometer for this purpose. (See the photograph of Dr. Bull with a large galvanometer of his design, fig. 48.) Hence he arranged a galvanometer with six strings, each string connected to a microphone. The microphones were placed in a linear arrangement more or less fronting the expected position of the enemy's guns and each microphone was considerable distance from its neighbors. A preliminary trial was made at the Marey Institute, early in the war in the presence of some of the French Generals. At this time they located the gun which was about five kilometers away to within five meters of line and ten meters of distance.

In the firing of a piece of artillery, there are two distinct sound waves,

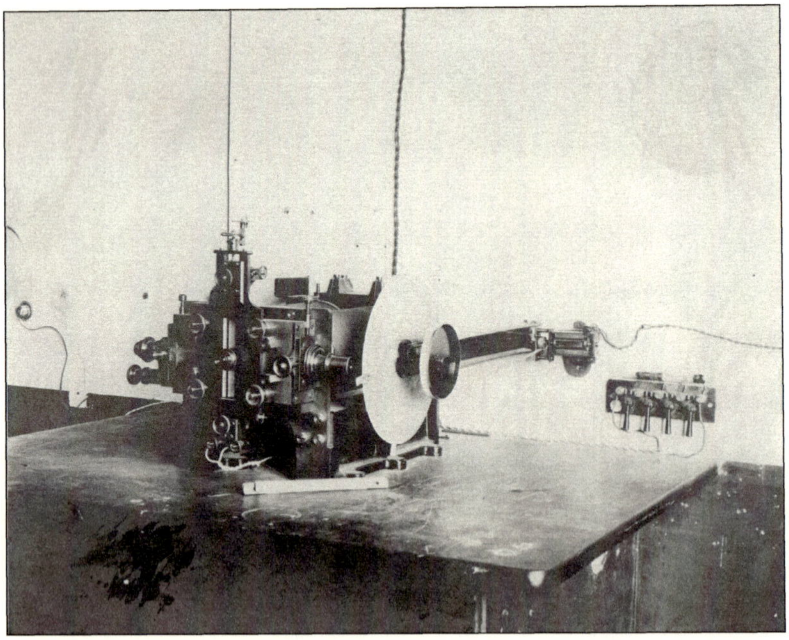

FRANCE

112

Fig. 48 — Dr. L. Bull making some adjustments on a large string galvanometer of his design at Marey Institute, Paris.

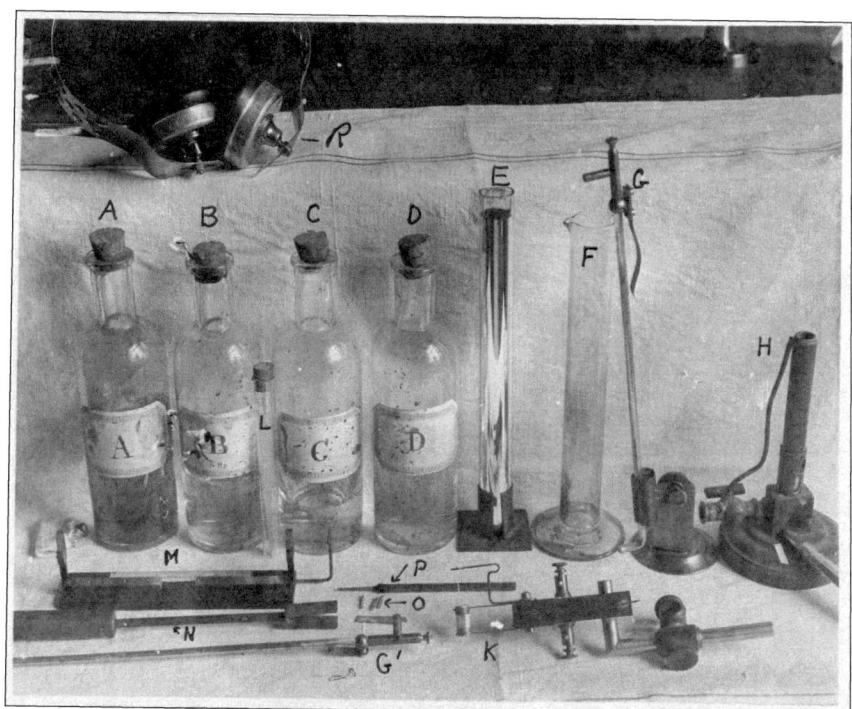

Fig. 50 — Details of apparatus used by Dr. Bull in drawing and silvering glass strings for the string galvanometer.

131

Fig. 49 – The Marey Monument in front of the Institute. Dr. Bull holding the first motion picture camera.

one made by the explosion of the gun (at the muzzle of the gun), the other by the flight of the shell. In the preliminary trial, a blank charge was fired, no shell passed through the air. It so happened that later, when they tried actually at the front, they had absolutely no success at first. It was because the sound waves produced by the shell flight entirely disturbed the record. Much time and effort were spent in developing a suitable microphone which would be sensitive to the wave from the muzzle of the gun, and not to the wave from the flight of the shell. This part of the apparatus was finally devised by a Mr. Tucker of England. In principle, his microphone is a box resonator with a hole in it and a glowing platinum wire or grid stretched across this hole. The wave created by the explosion at the muzzle of the gun is of large amplitude and causes a movement of air which is sufficient to cool this glowing platinum wire and produce a quick current change causing a deflection in the galvanometer. The string galvanometer, a lamp, camera, switch-board, etc. were all mounted on a panel about a meter square, and the whole thing was in the control of a scout. This scout had a position in front of the microphones so that when he himself heard the sound wave, he could, by pressing a button, set the galvanometer and everything in operation. He waited until the sound wave had swept past the microphones and released the button. The attendant in the motor car with the apparatus had only to cut off the film, develop it and hand it over to those in charge of the gun fire. Bull said the gun could be located as accurately as in their first trials and within three minutes of its discharge.

The British Army had such an outfit located every three miles on their front. They used no other apparatus for this service. The French used this arrangement and also one other type. I understand the Germans did not employ such a galvanometer arrangement, but recognized fully the effectiveness of this one, and issued orders that two batteries were to fire simultaneously with a view to confusing the enemy in locating guns. It so happened that this did not seriously disturb the localization. While perhaps the sound waves from the two guns would arrive at one microphone at the same instant, making it so that the record from that microphone was worthless, the other five microphones would give sufficient data to locate both guns. I learned that Mr. Tucker had received L.5000 for his microphone from the British Government. The Tucker Microphone was patented, but Bull's arrangement was not. He had made a claim to the British for financial remuneration for inventing the method and perfecting the rest of the apparatus.

Quite a number of the string galvanometers and parts of the apparatus were manufactured in the machine shop at the Marey Institute during the war. The profit of this enterprise is now what keeps the Marey Institute going. Their regular income is 25,000 francs, the same as before the war. They have five people working on full time. With their profits from the manufacture of instruments, they will be able to carry on for about eighteen months, at the end of which time, they will be at a financial crisis as an Institute. Bull, as assistant director, is practically entirely responsible for the work of the Marey Institute at present. He informed me that all the chief governments of the world had sent them work of some character with the exception of the United States and the British Government. Bull told me that the British Government is now disposing of these sound-rangeing outfots and he advised, if possible, that we secure one of them, as the apparatus would undoubtedly be useful at some time or in parts.

Another large research, which Bull has on hand and which grew out of the war is the matter of photographing the flight of a bullet as it leaves the gun. He has developed a technique for taking 20,000 to 50,000 pictures per second of the bullet as it passes from the gun. He has arranged so that a large electrical condenser discharges through a very small condenser. When the charging of the small condenser has reached a certain tension, it discharges by a spark across the terminals then the condenser is again charged and discharged over and over with a rapidity of such values as mentioned; i.e., 20,000 to 50,000 times a second. It is, I think, called the "bucket" method of discharge. A strong current of air must be blowing between the terminals so as to sweep away any luminescent particles that might remain there. Bull tried to measure the duration of one of these discharged sparks by an arrangement of revolving mirrors. In place of moving the film he, by the revolving mirror, switched the beam of light and got 5000 meters per second of movement. Even with this terrific speed, the image of the electric spark is, he said, entirely sharp, indicating that it has not been distorted by this movement of the light beam. So far as he has been able to prove, the spark is absolutely instantaneous. I can not understand how it is possible.

Bull has devised a special camera for working in this research. The wheel that carries the film is 16 cm. in diameter. Both edges of the film thread into slots so that it will not fly off the wheel. The speed of revolution causes the film to move at a rate of 300 feet per second. The slit of the camera, the muzzle of the gun and the discharge points of the condenser are all at the same level. The recoil of the gun closes a key and causes the large condenser to begin to discharge. The latency for the condenser discharge is about the same as the time duration from the moment the bullet leaves one end of the barrel and gets out of the other end. The bullet passes between the sparks and the camera, when out of range it cuts a fuse wire and thus breaking the circuit stops the condenser discharge, the picture is therefore its shadow in various positions. The picture is so sharp that the individual grains of powder which come out of the muzzle of the gun, unburned, are exceedingly clear cut. It is the burning of these unexploded grains of powder which makes the flash at the muzzle of the gun. Bull was greatly interested in some account of the apparatus, arranged by Mr. St. Claire of the General Electric Company, Lynn, for measuring the time of the bullet's passing through the gun barrel. He would like for me to send him any details that Mr. St. Claire is willing to supply regarding this apparatus.

The work was in progress with a common army rifle at the time of my visit.

He had a simple sandbox at one side of the room in which the bullet imbedded itself after being fired. Previously, he had been working with small field artillery, I believe, a three inch gun. He had a temporary building in front and to the left of the Marey Institute and a very substantial gun pit made of stone work which is located rather near the Marey Monument. (See fig. 49, stone work in the lower left hand corner.)

Bull very kindly demonstrated for me his method of making and silvering glass fibres for the string galvanometer. I remember he did some work along this line while he was visiting Nutrition Laboratory in 1913. Figure 50 shows all of the apparatus which he required for this technique. The various items are as follows: (A) solution of distilled water and nitrate of silver, the "Fondu form", i.e., not crystallized. An 8°/₀ solution is used. (B) solution of distilled water and pure ammonia. A 20°/₀ solution is used. (C) solution of distilled water and pure caustic soda taken from sodium, 3.4°/₀ solution. (D) a solution in the proportion of 25cc. of distilled water, 7.8 grams of white granulated sugar and 3.125cc. of nitric acid boiled together for two monutes. When cool, add 2.5cc. of 90°/₀ ethyl alcohol. (E) is a tube in which to do the silvering of the strings. (F) a graduate, for measuring the solutions. (G) a holder for the glass fibres to conveniently manipulate them and place them in the silvering tube E. (G) a second holder for manipulation purposes. (H) a Bunsen burner with a pilot light, very convenient. (K) the small electric furnace in which the glass rods are melted so as to draw them out into a thin string. (L) a tube of small glass rods ready to be used to make strings. (M) a holder in which withhthe help of the soldering iron (N), to solder the strings to the small pins (O). (P) a small polished boxwood stick polishing the strings after they have been silvered and have become dry . (R) a telephone set which, in combination with a dry cell, is used to test the strings for conductivity.

The small glass rods are of ordinary glass, about the diameter of a pin and four inches long. One end of the rod is melted, so that a small ball is formed. The electric furnace K, although not clamped on a support, is shown in the proper position. The arm at the top is slotted so as to catch this ball made on the end of a glass rod and to support the rod so that it will hang free inside the electric furnace. The rod extends about two inches below the furnace. A small hook is made on the rod by melting and turning the glass and on this a little bit of copper wire or some small weight is suspended. The electric furnace is supported about five feet from the floor. The current is turned on, one watches the glass rod and endeavors to turn off the electric current at such a time that the glass will have elongated to about four feet. The lower part, bearing the small weight of copper should not bump against the floor. A bit of shellac, touched to the upper horizontal bar or G and then touched to the string immediately below the electric furnace, allows one to break off the string and to carry it about. The string does not go floating through the air since it is taught by the weight hanging below. Then one winds the string over the lower horizontal bar of G and brings it back to the upper one, thus having provided two strings of proper length for silvering and both on same frame.

To silver the strings, take one volume of A and add one volume of B, then shake until clear. Add one volume of C and shake again. Pour this out of F into the tube E. Then, pour into E one volume of B and immediately gently immerse the string holder G with the two strings in proper position for silvering. The glass strings cannot be cleaned preparatory to silvering as a flat mirror should be, but they must be silvered very shortly after they have been drawn out, not the next day. The clip on the back of the silvering frame is helpful in holding the frame in the tube without tilting over and breaking the strings. Allow the strings to remain in the vessel E until the silver is well deposited on the walls of the vessel, and then take them out. The strings dry almost immediately. Polish them with such a piece as P for about five minutes. The polishing is needed only in the middle section which will show in front of the microscope after the strings have been mounted in the galvanometer.

When the strings are to be soldered to the pins which mount in the galvanometer, a very small piece of paper is touched with shellac and fastened to the lower end of one of the strings while the string holder is held in an upright position as G in the photograph. Then the string is cut below the point of attaching the paper. This makes a small weight which holds the end of the string down and does not allow it to fly about. On the frame M, which holds the pins, when the string is to be attached, there is at the extreme right, a small moveable arm. By a bit of shellac, the upper end of the string is attached to this moveable arm. Then one can, using the weight of the paper on the other end of the string, manipulate so as to lay the string across the two pins to which it is to be soldered. A small amount of "Woods metal" solder is placed on top the string at each point and the soldering copper N is warmed. The copper is not placed in contact with the solder but, since it has a slot in it, is slipped around the supports on which the pins are clamped. This warms up the support and the pin and thus causes the solder to melt without the danger to the string of directly touching it.

The clamps which support the two pins are insulated from each other and by using the telephone receiver in series with one dry cell and attaching one side of the circuit to one of the clamps and touching the other alternately, one can easily test by ear whether the string is electrically conducting. It is Bull's opinion that usually the strings for the galvanometer do not need to be made according to the method of William, by bombarding the strings with silver or gold particles in a vacuum. It is not necessary for them to be so smooth as this method provides. The air damping which is resultant from the strings being somewhat rougher as is the case when it is silvered by chemical deposit is not enough to introduce an error in the usual use of the galvanometer and Bull thought it was quite terrible to think of paying as much as $20.00 or $25.00 for a string.

Bull had a number of mechanical and optical things that interested me greatly. One of these was a recent addition to his splendid high speed camera for use with the string galvanometer. This camera is made on the principle that a piece of film is wrapped around a drum and this drum is given a heigh speed. When the film which has a length corresponding to the circumference of the drum has been exposed, the drum must be taken out and the film changed. Bull desired to take a long record of a spiral nature on this single piece of film. He, therefore, needed to arrange a small opening in front of the slit and have this opening gradually pass from one end of the slit to the other end and thus expose the whole film spirally and this required that the image of the string should all the time be kept on the moving opening. Immediately in front of the small opening which travels across the length of the slit, he mounted a small right-angle prism. A second prism he mounted at the end of the slit in such a way that it does not move. The image of the string from the galvanometer falls on this second prism and is reflected from it to the other prism which moves across the field and from this prism is reflected to the film. This arrangement Bull calls an "optical train". This allowed Bull to take very long high speed records in standardizing some tuning fork for a certain commercial company.

There was a fine magnetic clock made by a concern in Paris, which will operate from two to six years without adjustment or attention. Before the war they cost about $20.00. Bull has arranged that the pendulum of this clock carries a small convex mirror and immediately in front of the lowest point of the pendulum swing, a small magnifying glass is located. The clock is at the end of the room above his camera. A tiny pin-hole in the house of his arc lamp passes a beam of light so as to fall upon the convex mirror carried on the pendulum. This beam of light, as it passes the lens, on the frame of the clock, is reflected to the slit of his camera and so gives him an accurate time line for standardizing purposes. He has a very small, simple wireless receiving set by which he can get the wireless time signals and assure himself from week to week that the clock is accurate. He finds these arrangements very convenient for standardizing purposes.

Their storage battery at the Marey Institute is made up of 36 two volt cells, eached is wired to a pair of mercury cups in a horizantal board. By connecting up these mercury cups they can secure six, twelve, or seventy-two volts in the wiring over the building. The connection is most simple. There are several boards very clearly marked which have different arrangements of contact points, and by laying one of the boards upon the switch board, these contact points dip into the mercury cups and provide automatically for the voltage which is designated on the board. Any boy can go to the battery room and change "boards" to give the voltage wished by the experimenter, the arrangement is fool proof.

The machine shop at the Marey Institute is equipped only with foot-power machines, which was a surprise to me.

At Dr. Bull's residence, I had an opportunity of seeing the collection of stereoscopic views which he took at the Nutrition Laboratory in 1913. These views were mostly of calorimeter number three, which he was at that time investigating with a view to duplicating that apparatus in Paris for the Societe Scientifiquw. These steriopticon photographs were on glass and were viewed by transmitted light. I think I never saw any pictures which more accurately portrayed apparatus. Such pictures would be invaluable as an accompaniment to diagrams if one were trying to duplicate an intricate apparatus. Their shief merit is in making clear the perspective especially when viewed by transmitted light. In this regard they are, I think, just as good as an actual small model which is meant for viewing rather than for mechanical operation.

Dr. Nogues is continuing the work with high speed motion pictures of human subjects. He had lately arranged a camera which he showed me. It outwardly looked very much like any motion picture camera but would take pictures at the rate of 300 per second, in open sunlight. The picture taken of athletic events with this camera and projected at the normal rate are highly instructive in regard to the mechanics of physical motion. He has, I believe, gotten his camera so that it is a commercial proposition and can be used for certain scientific investigations in industry or for matters of providing films for entertainment. His latest work is to produce stereoscopic motion pictures and the last day of the Physiological Congress he asked me if I could conveniently come out to the Marey Institute and see a demonstration which he was going to give. I did not learn full details but believed a great deal of the stereoscopic effects is produced by the specially designed screen on which the pictures are projected. Dr. Nogues has a small machine shop in one of the upper rooms of the Institute and does most of his own machine work. One of his special interests and lines of research is the shape of air and water propellers. Dr. Nogues is a doctor of medicine.

I have gone into considerable detail concerning my visit at Marey Institute. It is a place where details are all most instructive. Also it is a purely research institute and hence of special interest.

122

Physiological Laboratory of the Sorbonne.
Professor Bapicque and Dr. Stodel.

Professor and Madame Bapicque I found at their laboratory busy with an investigation on a problem of nutrition which had come to them during the war. It was the utilization of various kinds of sea moss and seaweed as food for animals. They were experimenting with feeding chickens. The chickens ate the sea weed very readily. There were two varieties which looked similar but did not taste alike. It was quite impossible to fool the chickens. They would not even make one peck at the poorer variety. Lapicque said he could subsitute a certain variety of sea weed for the oats in the diet of a horse and the horse would eat the seaweed with practically equal relish but it was not so satisfactory if the horse was doing hard work.

Bapicque was interested in discussing with me certain problems of nerve physiology which had to do with electrical stimulation. He has long worked on this field of electrical stimulation of nerve and muscles for threshold contraction and is one of the chief leaders in this branch of physiology. I showed him the nature of the apparatus which I had been using for threshold stimulation of the finger tips in man. He gave me a reference Sepshenaw 1860, "Summation of Nerve Irritants". He discussed means of varying the stimuli as to number, frequency, duration and intensity of the electric shock and told me of variations in the use of such stimuli to illicit reflexes such as the vasomotor inhibition, as in the vagus secretions and pigment motivation. He discussed the curve which one usually obtained when plotting intensity against time in relation to the electric shock which just produced threshold contraction or sensation.

Bapicque seemed somewhat depressed. (A reasonably good picture of him is shown in figure 51). During the first year or so of the war he served as a

Fig. 51 – Professor Louis Lapicque with apparatus for determining the "chronaxie" of nerve in the Physiological Laboratory, Sorbonne, Paris.

Fig. 53 – Ph. Stodel in his laboratory for Physiological Chemistry, Sorbonne, Paris.

WALTER MILES AND HIS 1920 GRAND TOUR

Fig. 56 - The balcony above the court was a favorite promenade for Congress members. Here W. R. M. took some "snap-shots", Figs. 57, 59, 60

Fig. 58 - Prof. Langlois (left) demonstrating to Congress.

I read my paper on the Pursuitmeter in this same session and room.

Fig. 59 - Prof. Richet (left) Pres. of Congress, discussing politics.

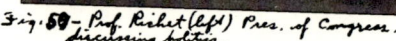

144

surgeon in the regular army. It proved that he was physically unable to keep this post and he took charge of some work in reference to nutritional problems. He tried to advance by propaganda in France the use of more dark whole-wheat bread. He had quite a political task in securing the election as Professor of Physiology at the Sorbonne in that place made vacant by the death of Professor Dastre. (Dastre was on day talking with a friend about his own health and he said "I am so well I have no idea of what disease I shall die." Within a week he was struck and killed by a military motor car. Dastre's daughter had used much artistic skill in decorating the walls of the vestibule of the Physiological Laboratory, see figure 52. Lapicque did not like the large painting representing the "Beginning of Vaccination" on the wall of the vestibule and during the Congress had the painting by Thermitte representing "Claude Bernard Demonstrating an Experiment to His Friends" from the Luxembourg Gallery placed there). Lapicque was previously in the Museum of Natural History associated with Dr. Tossot, and claims that much of the apparatus which he built and used at the Museum of Natural History was his own personal property. Now that he is in his new position, he desired his apparatus, but Dr. Tissot will not let him take any of it from the Museum of Natural History, claiming that it belongs to the Institution. Lapicque feels that so far as his scientific work is concerned he has lost the five best years of his life and that he now finds himself in a new position without apparatus and with resources very meager for the equipment of his laboratory. He does have a number of pieces of apparatus however, among these I saw the Lucas Pendulum Circuit Breaker. It was the only place where I saw this apparatus outside of Adrian's laboratory. Lapicque and his associate, Dr. Stodel, were very busy making necessary arrangements for the convenience of the Congress of Physiology which met in their laboratory using its various rooms for demonstrations and their lecture amphitheater as well as three other adjoining amphitheaters for simultaneous scientific meetings. Stodel (see fig 53) was present during part of my conference with Lapicque but I did not see his laboratory. His work is chiefly chemical. I met Professor Gley here also, he was entirely occupied with matters of the Congress and spending no time in his own laboratory across the street.

Psychological Laboratory of the Sorbonne.
Professor H. Pieron.

Professor and Madame Pieron were working at the Psychological Laboratory on a visual problem. They had quite a complicated apparatus for presenting the visual stimuli varying them in strength, in duration, and also as to wave length. The duration was manipulated by a large rotating disc with slits which were variable in extent. He was determining the sensitivity of different portions of the retina to different qualities of light stimulation.

A research student, Mr. D. Wechsler, a man who had worked in the psychological laboratory at Columbia University with Professor Woodworth, was investigating emotional situations. It was his ambition in this research to arrange a scale of emotional situations which could be used in measuring the <u>feeling side</u> or development of personality as the common intelligence scale is used in measuring the intelligence of the individual. It seemed to me a pretty large order with small possibilities of attaining anything of very exact nature in results. Mr. Wechsler placed the hand of his subject in a plaster cast arrangement for recording the pulse rate. He used a pneumograph around the subject's chest for respiration rate, and was recording the respiration ratios. He was also considering the use of the psychogalvanic reflex measurement in connection with his problem. Mr. Wechsler kindly showed me about the place and he gave me a concave mirror arrangement by which one can obtain stereoscopic effect with one flat picture. A second research student was working on the problems of choice reaction, using visual, auditory and touch stimulation. There seemed to me nothing especially noteworthy in this research. Pieron had developed an apparatus which gave a great variety of speeds of rotation to color wheels. This gradation in speed was obtained by the use of two truncated cones with a small wheel acting between them.

I was much attracted to Pieron (See his photograph in fig. 54). He spends

Fig. 54 - Professor Henri Piéron at his desk in the Psychological Laboratory, Sorbonne, Paris.

Fig. 55 - Group at Société Scientifique d'Hygiène Alimentaire in Alquier's apartment, Paris.
Sitting: Mme. Alquier and Alquier
Standing: Alquier (son), Bertrand, Lefevre, Lemoine, and Compton.

his entire time in research and is the successor of Professor Binet. I understood there was a possibility of their organizing an Institute of Psychology at Paris to include the department of Pieron, Janet and Dumas. Then Pieron will likely do some teaching.

Pasteur Institute.
Dr. Bertrand.

Dr. Bertrand asked me particularly about the alcohol researches at the Nutrition Laboratory. He wanted to know if I myself used alcohol and did not think that a small amount of alcohol might have effects in the nature of freeing one from inhibition and making one more comfortable which would at times outweigh in importance any small decrease in one's production efficiency. He thought that the laboratory results were only one line of light on this question and there were other view points which represented parts of the truth in reference to alcohol and should be considered. To him it was quite beyond understanding how the United States could adopt prohibition and he wondered if this legislation would be reversed.

Dr. Bertrand and his assistant, Dr. Compton, showed me the chemical laboratory. They were at that time working on a problem of fermentation in vegetables canned by what is known as the "cold pack" method. Sometimes vegetables so canned ferment and, at other times they do not or perhaps I should say other cans do not. Bertrand thought that to discover the reason for this was a very nice problem.

I informed Bertrand of the special interest in America in industrial diseases particularly in poisoning which resulted from industrial employment. I told him of the establishment of the Schools for Public Health at Harvard University, among others and also of the work of Dr. Drinker on the subject of Manganese poisoning. In this subject I knew that he was himself much interested.

I believe that in Bertrand we have an example of the very highest type of scientific Frenchman (See fig. 55). It is greatly to be regretted that there are those of his fellow scientists who have blocked his election to the National Academy which honor he certainly deserves, by virtue of a long period of splendid scientific activity. Compton, his assistant, told me about the fact that he had for the present missed the opportunity of becoming a member of the National Academy, but that this had not soured Bertrand or altered in any way his scientific work. I felt that Bertrand was particularly considerate of my visit and although he was heavily burdened with certain committee matters he still came to the dinner at the house of M. Alquier and appeared to be greatly interested in their plans for building a calorimeter, although he said very little.

132

A dugout visited by our party.   Photo by Dr. W. W. Speakman.

152

FRANCE

Societe Scientifique d'Hygiene Alimentaire.

M. Alquier and Dr. Lefevre.

After arriving in Paris, I wrote a letter to M. Alquier asking for an appointment and to visit their Institution. There was some delay in receiving a reply to this request and one day thinking that the laboratory would probably be open and someone would be working there, and also that possibly Alquier might be out of the city, I went to the address given on the letter head. I found a splendid corner building inscribed with the name of the societe. The front door was slightly ajar and after knocking and receiving no reply, I looked in and found the place entirely vacant. It seemed to me at first that probably it had been occupied by some war work organization and lately vacated so that apparatus had not been re-installed.

Later I received a reply from Alquier which gave the same address but I found there was another entrance and another wing of the building which I had not previously noticed. In this Alquier and family have a living apartment which belongs to the Institution. A reception room, office and library are on the first floor. He told me that they had not yet built their calorimeter that during the war they had everything else to think of aside from building apparatus; that they were interested in food problems and particularly in helping with food propaganda in France. He said nothing whatsoever about the mission on which they had sent Bull nor about any report from him. However, he told me of the project which they had on foot to build a calorimeter according to the plans of Dr. Lefevre. He thought these plans to be very excellent and to embody certain decided improvements which would make their instrument the best of this sort which had ever been constructed. His chief interest, at that moment, was in securing the co-operation of Dr. Benedict and the Nutrition Laboratory in the construction of certain parts of the

FRANCE

Wm. Wolfram

Paris
17th Seventh Mo, 1920

Paris Medical School on left; Laboratories on right.

respiration apparatus which they could embody in their respiration calorimeter. He felt that it was quite impossible to have this apparatus construction done in their own country because there was no one who was especially adapted to do it, by experience in making this type of apparatus. I assured him of my belief that Dr. Benedict would be willing to do-operate to the extent of our ability.

In reference to publications from this laboratory sent to them, I asked at Dr. Benedict's request if it was their desire that publications should be sent to the library of their Institution, to Alquier and also to Lefevre. Alquier said that it was quite sufficient if one copy of each publication was sent to the library of their institution and that here it would be available for all of their use. He remarked that they were thoroughly interested in the publications of the Nutrition Laboratory and urged if possible they receive copies of each publication as issued.

Lefevre lives quite far out from the city of Paris. However, it was the kindness of Alquier to arrange a dinner at the Alquier apartment at which Lefevre, Bertrand, and his assistant, Compton, and also Dr. Lemoine, a friend of Lefevre's were present with the Alquier family. (See fig. 55). At this time I had an opportunity of seeing the blue print plans of the calorimeter which they proposed to build. These diagrams, together with a description of the apparatus, which they proposed, has already been published in the bulletin of their institute. Again, they were especially anxious to inquire if they could secure from this Laboratory a "universal apparatus" and certain other parts of out regular equipment which they wished to use as parts of their own. (This whole matter has been taken up very fully in correspondence between Dr. Benedict and M. Alquier). Dr. Benedict has criticised them for pub- a description of their apparatus before it has actually been constructed. I found out from Alquier that they had taken this criticism in a kindly spirit and were not disgruntled towards the Nutrition Laboratory. They showed me no apparatus whatever and I believe they have nothing in the form of a laboratory

I was bold enough to suggest at the time of discussing their plans following Alquier's dinner that they secure some simple respiration apparatus, i.e., the Benedict Portable respiration apparatus or the Benedict cot chamber apparatus and experiment with this on some of the problems in which they were interested. Thus they could get together something of a laboratory force and have experimentation in progress while their plans for their calorimeter were more fully developed. I explained that the calorimeters were not being used for much now in the Nutrition Laboratory since it was found that the determinationw obtained with respiration measurements were practically the same as the more tedious and expensive technique of using the calorimeter. They, particularly Lefevre, immediately objected that "The honor of France demanded that they have to begin with the most perfect apparatus possible". It was my impression that Alquier rather favored making a start at actual experimentation with some simple forms of apparatus. I was unable to determine just the attitude of Bertrand in the matter. He seemed to be a member of their board of trustees. There certainly are some who have not been pleased (e.g., Lapicque) with the quiescent attitude of this institution. They feel that it has accomplished nothing except to provide a position for Alquier and that during the war it hindered rather than helped, especially in reference to the bread situation since this institution insisted through its bulletins that only white bread was satisfactory for food. At the time of Alquier's dinner I was asked an opinion relative to the food value of white bread and dark bread. Before I could formulate an answer the question happily was lost and forgotten in the zest of conversation which followed a remark made by Madame Alquier.

From the slight amount of observation which was possible for me it almost seems that this Institution up to the present represents a scientific loss rather than an asset to France. Its endowment was raised and its splendid building erected by public subscription. They aroused public sentiment and enthusiasm for scientific research in reference to the very important problems of food and then nothing more of a tangible nature is accomplished. They occupy this field of inquiry so far as France is concerned so that no other organization or individual feels especially called upon to undertake the problems. They occupy the attention and time of scientific men with the discussion of their questions and projects. They spend their money sending Bull to America to gather data and plan for them and never asked him to report for a period of six months following his return and before the declaration of war. Although scientific work many times requires outlay with little of immediate tangible result it does seem that in the case of the Societe Scientific dS Hygiene Alimentaire up to the present a debt has been accumulating which will require a large amount of scientific activity and output to counterbalance.

Physiological Congress of 1920-July 16-20.

(Germans and Austrians excluded).

As I did most of my visiting of scientific laboratories before the Congress met in Paris, I naturally spoke of it many times and heard considerable comment. I think that especially the Dutch, e.g., Van Leeuwen, Einthoven, and Hamburger, were particularly interested in what they called "The French Congress." They seemed to think it rather unfortunate that the Germans and Austrians were excluded, suggesting that they might have been invited but that they probably would not have attended. This attitude on the part of the Allies and Neutrals would have promoted the beginnings of International Intercourse. After 1913 the next Congress was to have been presided over by Professor Tigerstedt and he had not been consulted at all in reference to the Congress, in Paris which had been called without any authority from the old International Committee.

I learned that the Germans and Austrians, inviting physiologists from neutral countries, held a physiological meeting among themselves in Hamburg, Germany, in June, at the newly built but still unfinished and unequipped physiological laboratory of Professor Kestner. Van Leeuwen, Einthoven, and Johansson among others attended this meeting. I understood that the Germans looked upon the proposed meeting in Paris as something that had been pushed through by the French and was another indication of the French attitude to crush out the Germans. It was reported to me that although Professor Rubner attended this meeting at Hamburg, he was given practically no place in that meeting, not being asked to occupy the chair or to address the meeting. My informant said that the German physiologists felt that Rubner had betrayed them by his attitude in the war, e.g., he was asked if he did not know long before the armistice that the Central Powers would be unable to continue from the standpoint of food supply. Rubner told the questioner that of course he knew the Central Powers

FRANCE

140

161

could not continue and that it was hopeless for them to go on from the stand point of such material.. When questioned as to why he continued to publish statements saying that the food supply of Germany was adequate to the needs and they could continue, Rubner replied that he published these statements which he knew to be false because the military authorities asked him to do so. I understood that many of the Germans feel that this action on the part of Rubner was unpardonable, that he should have given the Government advice as to actual condition. It is easy to look backward and say what others should have done.

The Dutch Physiologists felt that the next Congress after 1920 might best be held in America. They seemed to think that the idea of going to America for a scientific meeting would appeal to all of the scientific men of Europe. They spoke of the high cost of travel and several times it was mentioned if it would be possible for the Americans to supply a ship for the transportation. Even though one points out the fact that the Americans for years past have stood the expense of making trips to Europe for scientific purposes still they seem to feel that the Americans have at their disposal, immense wealth and that these expenses do not mean so much to us as they do to the European. They are not in habit of financing for themselves such long trips.

The list of members of the Paris Congress shows a total enrollment of four hundred and one. Since it is a common thing for a man to become a member of such a scientific meeting and send in a title for a place on the program without attending in person, I imagine that not more than three hundred of these people were in actual attendance. The meeting was well attended especially by French and English. From the United States the chief representatives of physiology were Lee, L. Henderson, Lusk, Flexner, Stewart (G. N.), Able, Hess, and E. Cohn. The President of the Congress was Professor Richet. (See fig. 59). The secretary was Professor Gley and the Treasurer, Dr. Bull. The opening address was by Professor G. Fano on the subject, "Inhibition and

Will." The list of demonstrations and communications together with abstracts of these has been printed in a volume which is deposited in the library at the Nutrition Laboratory. Of especial interest to the Nutrition Laboratory was the communication by Professor Noyons describing his new differential calorimeter. At the time of my visit to his laboratory he was working on a small model of differential calorimeter suitable in size for dogs and smaller animals. I suggested to him it would be first rate to demonstrate to the Congress. It seems that he took my suggestion to heart and made haste to complete this instrument and shipped it, with a great deal of trouble to himself because of its size which made it difficult to load on a common express car of the Belgian railway. He, therefore, had it in working demonstration at the Congress and it was a center of great interest particularly for M. Alquier and Dr. Lefevre.

I read a short paper on "The pursuit meter: An apparatus for measuring the adequacy of neuro-muscular coordination," as scheduled on the program of July 17, in the Amphitheater of Geology. There is really little satisfaction either in giving or listening to papers at these scientific meetings. At the same hour scientific programs are in session in four adjoining lecture Amphitheaters and there is a great amount of confusion occasioned by changing of individuals from one lecture room to another with the hope of hearing a certain paper or listening to particular friends read papers. Certain individuals have with them assistants who are translating audibly while the lecturer is speaking and if it so happens that some one desired to put up or take down charts or re-arrange slides or demonstration material at the same time that the speaker is trying to present his subject, it becomes such a confusion as to be almost hopeless. At the beginning part of each program it is usually quite satisfactory. The Congress is unique in offering an opportunity to see a lot of men whom one would not be able to visit personally but for those who are working in the same field or in whom one is particularly interested, it is much more satisfactory to see them at their homes and in their own laboratories.

Figures 56 to 60 inclusive show views at the Sorbonne Physiological Laboratory and at the time of the Congress.

The social occasions in connection with the Congress included a reception at the home of Professor Richet, an excursion on Sunday, July 18, to Chateau de Chantilly, a program of scientific motion pictures followed by refreshments and a reception to the members of the Congress by the Recteur of the University in the Salons of the Sorbonne.

Concerning the meeting of the next Congress some of the Americans felt that it was not wise to invite it to the United States due to the expense and the distance and that particularly now with the depreciation in the value of European money, which may continue for some years to come, the Congress would not be a success in America. They felt also in reference to the matter of inviting the physiologists of the Central Powers to meet with them that it was only right for the Americans to follow the lead of French and English on this point and since they have suffered so much more than the people in the United Sataes, it was hardly becoming of us to insist on immediately re-establishing relation.

When in Germany, I spoke with Professor Kestner and Professor Rubner asking what was their opinion about the possibility of again holding truly international scientific meetings. Both of these men answered me in practically the same words which were in substance as follows:"With the Americans, of course, we can meet and"(there was a little hesitation,"also with the English, but with the French, never!"

Dr. and Mrs. Waller came to the Conggress in their automobile and afterwards drove from Paris up into Belgium and visited with the family of Miss de Decker their assistant. Sherrington and Lee (the latter is staying in England for a year working in connection with the Industrial Fatigue Board) went from Paris to London by aeroplane. There were no general excursions following the Congress as was the case in Groningen in 1913. It was proposed to hold the next

Congress in 1923 and the invitation of Professor Schafer and his colleague, Professor Cushney, was accepted for Edinburgh. Lusk was chairman at one of the sessions. He had expected to go to Vienna following the Congress but changed his plan on account of something said by Dr. Field, who is working on the International Bibliography of Scientific Literature. He told Lusk that in Vienna at that time they were getting more food than previously and conditions were much better. Dr. Field was earnestly soliciting financial assistance for his Scientific Bibliography.

Dr. Jean Le Goff, a Paris physician, (178 Faubourg St., Honoré VIIIe) has been interested in The Carnegie Institution of Washington and particularly in the Nutrition Laboratory. He has written a short description of the Institution for a French journal, furthermore he kindly sent us some rare books. I had the pleasure of seeing him and taking tea at his residence. His wife is an American lady from New York. Our discussion was in reference to alcohol effects, Prohibition in the United States, and the possible connection (he thought it certain) between the candy habit and increased diabetes. He has an American examination chair in his office and tells you proudly that it is the only one in Paris.

At Chateau de Chantilly I first met Professor Maurice Nicloux who has published much on the alcohol question. He is now located at the University of Strasbourg.

Belgium

145

Solvay Institute, Physiological Laboratory.
Professor Heger and Professor Phillipson.

Due to a mistake on my part, I went to Professor Heger's residence, thinking it was the address of the Physiological Laboratory of the Solvay Institute. This occasioned some delay, it was also a holiday and Heger had kindly remained during the morning to visit with me and show me the Physiological Laboratory. He had an engagement for the afternoon hence my time with him was somewhat limited. It also happened that Professor Phillipson was away on a vacation. He had poor health due to a term of imprisonment which was brought about during the war by the fact that his son entered the Belgian Army. Heger is no longer teaching Physiology, his professorship having taken by Phillipson. He is now chairman of their Board of Council for the University of Brussels and heavily loaded with administrative duties in connection with the University. He explained that the building of a new state railway station in the vicinity of some of their main university buildings was going to necessitate moving the University to another location which they have already selected. The planning of new buildings and moving of the University is requiring a tremendous amount of time and thought from Heger. He was interested in the plan of the Harvard Medical School as I happened to have one of the aeroplane views of the Harvard Medical School and adjoining buildings. At his request I was able to supply him with this photograph and later with a number of details relative to the ground plan of Harvard.

Heger showed me completely the physiological laboratory so far as the rooms and equipment are concerned. He knew nothing about the researches in progress. As is well known, the laboratory is a very splendid Institute of its kind having been thoroughly planned by Heger himself. The Museum in which are collected works and memorials in reference to the researches of Professor Stas is quite remarkable and worthy the attention of any scientific man. I was particularly intersted in the pair of balances made by Stass when he was a small boy compared with those which he later made and used in his remarkable researches in standardization. Heger showed me some well preserved physiological specimens which illustrated splendidly the functional activity of the membranes surrounding the intestines and the liver in removing certain poison which had been given to these animals from which the specimens were later taken. I had a very short conversation with Dr. E. Zunz in the department of physiological chemistry.

Heger had a great deal to say about the war and conditions there during the German occupation. He seemed to feel that the German scientific men during the war carried on such a propaganda of absolute misrepresentation with the objective of breaking the moral of the Belgian people that he never again could really believe statements made by these scientists. Research work in Brussels had been practically at a standstill since the war but I am sure it would have been well worth my while to have had an opportunity to visit with Phillipson and to have seen something of his work. Under the circumstances of the holiday and Heger's leaving the city in the afternoon I was fortunate in being able to visit the laboratory at all and particularly am I grateful to him for insisting that I see Professor Noyons and his new Institute for Physiology at Louvain.

147

Catholic University, Physiological Laboratory.
Professor Noyons.

Professor Noyons was for ten years an assistant to Professor Zwaardemaker in Utrecht. The assistancy here paid him very little and he made his living through the practise of medicine. A picture of Professor Noyons is shown in figure 61. In 1911, I believe, he was approached with regard to accepting the chair of Physiology at Louvain and after a year, made up his mind to accept that position and went there in 1912, during which year he built a very nice residence for himself. One of the conditions under which he came to Louvain was that he should be allowed to build a new physiological laboratory at that University. After a visit to other physiological laboratories, particularly that of Professor Hamburger in Groningen, he completed his plans for the new laboratory of Louvain and the foundation had been laid for the structure and the walls were up to about the first story when war broke out in 1914. Furthermore, he had purchased a good deal of apparatus or had the orders placed a and he had received the money from the University authorities for the purchase of these materials. As a physician he had been first assistant in the Louvain city hospital. With the beginning of war the head of the hospital went into the Belgian army and Noyons being a neutral (a Dutchman) and so having not to do military duty came into position as head of this hospital. He, therefore, was in the city of Louvain during all the war and as head of the hospital and a neutral professor in the University, occupied quite a strategic position. He perhaps has had the best opportunity of any one to gather unbiased date concerning what actually happened at Louvain during the German occupation of that city. He should, I think, through some agency, such as the Lowell Institute, be invited to give a course of lectures and so put this material in shape for historical record. I found him a man of very open mind without a spirit

Fig. 61 – Professor A. K. M. Noyons at his office desk in the Physiological Laboratory, University of Louvain.

Fig. 62 – Professor Noyons (right) and Dr. Libbrecht (next) with assistants in their shop. The two round objects are shell cases being made into respiration chamber windows.

Fig. 70 - Partial view of the new Physiological Institute, University of Louvain.

hate and I thought him quite unprejudiced in his general view point. He told me it was his opinion that the burning of Louvain might easily have started by some one discharging a gun accidently.

On two occasions, the soldiers tried to burn Noyon's residence. The first time he learned of it through his maid being brought wounded to the hospital. He then secured a body guard from the German Commandant of the City and going to his residence, compelled the soldiers, who had set it on fire to fight the fire and put it out. On the second occasion when an attempt was made to burn his home, no one was there and the fire smothered out by itself. He complained to the Commandant of these efforts and a notice was put on his house that any one, soldier or civilian, found in the house would be shot. This made it so that Noyons and his wife had to live elsewhere during the remainder of the war and were not in the house at all. Their residence overlooked the railway tracks along which the principle movements of German troops were carried in through Belguim. It was quite possible The Military did not desire to have any one living in the house and so in position to count the number of trains and cars loaded with troops that passed or were in that particular switchyard.

It was forbidden, of course, in Louvain to have copper, lead, glass or building materials of any kind. Without special permit, however, Noyons, without asking any questions, proceeded to continue his building of the Physiological Laboratory. Most of the other buildings in the neighborhood and most of the University buildings were by this time burned down. No man of military age or fitness remained in the city. He employed others, many of them not skilled in the things which he desired them to do, but he taught them how to lay bricks and how to proceed with his building. He superintended their plumbing, they built all their furniture, and in one way or another, secured and installed apparatus. This he did at times when he could leave the hospital. I think that the building of the Physiological Laboratory at Louvain during the war while the Germans were occupying that city and with the city practically in ruins (there were only nineteen houses standing on the longest

BELGIUM

150

Dieses Haus ist
zu Schützen
Es ist streng verboten, ohne
Genehmigung der Komman-
dantur, Häuser zu betreten
oder in Brand zu setzen.
Die Etappen-Kommandantur.

Fig. 71 - The electrocardiogram outfit and room at the Physiological Institute, Louvain. A duplicate of Salomonson's apparatus.

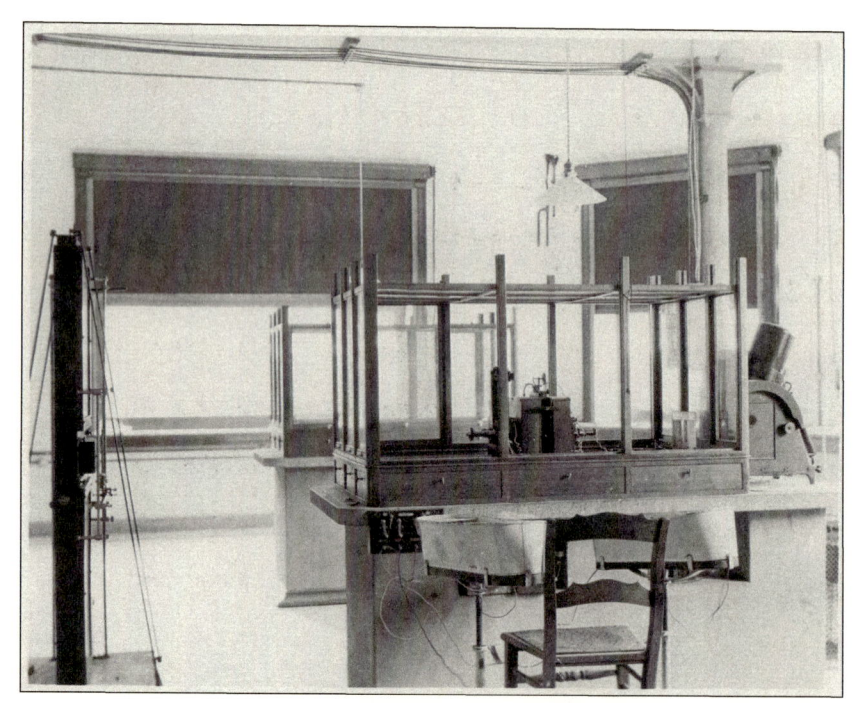

Fig. 72 - View of electrocardiogram outfit showing arm-electrodes, and a second pillar with case for instrument. This is a corner basement room. Windows fitted with special black shades, walls white. The pillars of cement. Top has groove around the edge so that small things will not roll off. The case about the galvanometer made of several glass slides which can be taken out leaving only top and frame work.

street in Louvain, a street some two kilometers in length) is one of the most remarkable achievements which came under my notice. Due to the war donditions and the fact that the laboratory had to make things rather than to buy them, this Institute is equipped in a very practical way and I think that Noyons has been extremely successful in taking advantage of the good points in the arrangement of other laboratories and in combining these and adding certain things of his own origination. He had a group of three young men with him, Dr. Libbrecht, instructor, and Mr. Fauville and Mr. Bouckert, all very keen and working tremendously to get things going and to promote the interest in physiology in Louvain. One wing of the building devoted to physiological chemistry was not yet occupied. The professor in this branch has just been selected, Professor F. Malengreau. On the attic floor of the Institute the books which had been given to the University Library were stored temporarily.

Noyons was having constructed in his shop at the time of my visit a small differential calorimeter for use with dogs and smaller animals. Parts of this are shown in figure 62 with the men. From left to right these man are: Fauville, a student; Bouchert, a student; Libbrecht, instructor; and Noyons, professor. Fauville is holding on his arm a German shell which has been cut off and a hole made through the bottom to adapt it for a window for a respiration chamber. The group are standing about a small metal box on the outside wall of which flat tubes have been soldered so that water can be circulated in them for controlling the inside temperature. The lid is standing down in front. A second similar chamber is shown behind Noyons. On one of these chambers the animal is placed while in the other electrical resistance wire is used to produce electrical heat to compensate in the galvanometer circuit the heat produced by the animal. The two chambers are mounted side by side so that their environmental condition will be as similar as possible. They are ventilated in the same way and the temperature of the water in the the walls of the chambers is as nearly the same in both as possible." Noyons Noyons brought this small calorimeter complete to the Paris Congress and demonstrated it there. In principle it is the same as his large calorimeter.

In connection with Noyons communication to the Congress in Paris he used four charts to explain his larger calorimeter. These four charts I have photographed and they are shown in figures 63, 64, 65 and 66. Figure 63 shows the ground plan of the differential calorimeter at Louvain. This is located in a large basement room of the Physiological Institute which room is devoted entirely to the calorimeter and was designed for that special purpose. Up to the present time, Noyons has not had an opportunity to fully work out his results with this calorimeter nor to describe the apparatus. Under these conditions it seems to me best that I should not make measurements and go into too many details in reference to the calorimeter at the time of my visit. I informed him that Dr. Benedict had made and had had in operation for some years a differential calorimeter for small animals, that he controlled the environmental temperature by blowing air over the chambers and used electric heat on one side to counterbalance animal heat on the other. I said nothing to him about his integrating meter arrangement. Figure 64 shows the schematic side elevation of the calorimeter. The floor of the basement room was especially designed and fillers located on which to build the two chambers X and Y. I judged that each chamber X and Y is about $2\frac{1}{2}$ meters long by $2\frac{1}{2}$ meters tall and $1\frac{3}{4}$ meters wide. This I mean to be clear open space not counting the space occupied by the pipes and by electric wires on the wall. The two chambers enclosed by a series of walls beginning at the extreme outer wall A (See figs. 63 and 64) we have two double wooden walls, B is an air space, C a cork wall, D a wooden wall, there being a dead air space between C and D, F air space and H the inside wooden wall to which is fastened the lead lining, I. On this lead lining secured by porcelain insulators are the water pipes K and on top of these a distance of three or four centimeters from the wall is a layer of wire gauze L. This This takes up the heat very readily and communicates it to the water pipes.

It was Noyons' original plan to measure the heat production of the animal organism by the change in the temperature of the water in the two chambers hence the apparatus was built to work on this principle. He later developed the idea of using electrical heat on one side to counterbalance animal heat on the other and now uses the water pipes only to control the environmental temperature. The air is kept moving in each chamber by the fans M and N which are driven from the same motor and by the same sort of belt system P so that their speed will be uniform. The doors opening into the two chambers from the hallway Z are exactly the same size and spaced similarly. They are large enough for a man to walk in by stooping a little. Both chambers must be lighted artificially. There is a glass in the door of each chamber and the light hangs equidistant in the hallway Z. It is arranged to be moved in the hallway but so that it cannot be placed nearer one chamber than the other. At present there is no arrangement for collecting the expired air. It is Noyons' plan to make it into a respiration calorimeter.

The subject enters the chamber on the side designated as A in figure 63. The outer shell of the calorimeter has in it certain openings whereby it may be ventilated. These are seen in figure 63 also as two square windows in figure 66 which is a photograph of a chart showing the arrangement for ventilating and regulating the temperature of the calorimeter. In two of the direct photographs, figures 68 and 69, these square windows appear. The control apparatus is on the rear side B in figure 63. This consists of an arrangement for measuring the temperature of the water in the pipes of the two calorimeter chambers and for regulating this temperature either by the use of an anomia refrigerating plant, a small one, which is shown under the bench and at the left hand side of figure 66 or by putting warm water into the pipes of the chamber. Noyons claims that he can make experiments, regulating the environmental temperature from that which would correspond almost to freezing to that which would be torrid. The ventilating pump is seen under the bench in the photograph of the chart figure 66 also in the direct photograph figure 67 where Noyons is shown in the picture balancing the electrical heat against the animal heat. In combination with the ventilating system they use a

Fig. 67— Professor Noyons balancing electrical heat against animal heat in his large differential calorimeter.

Fig. 69— Details of the differential calorimeter. A "Rotamesser" at the left. Mounted on and in the cabinet are the arrangements for regulating the water temperature in the two chambers.

Fig. 68 – The electrical control apparatus of the differential calorimeter.

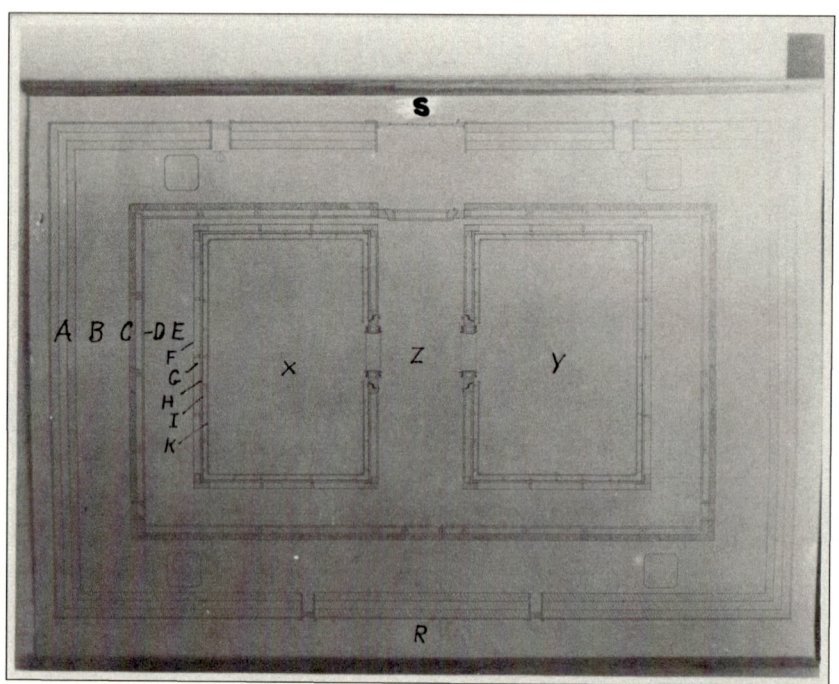

Fig. 63 – Photograph of a chart showing the ground plan of Professor Nayon's differential calorimeter, Louvain.

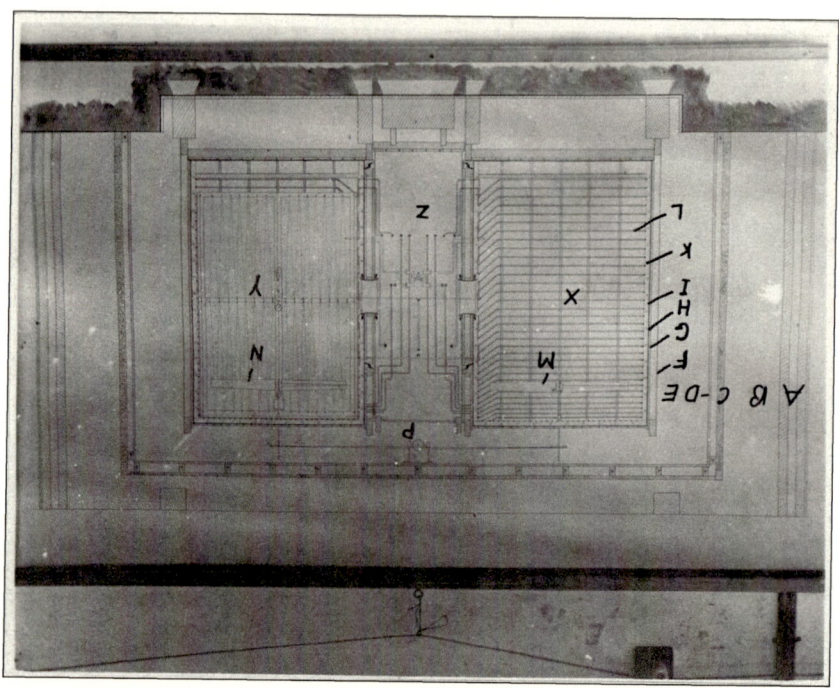

Fig. 64 – Photograph of a chart showing the schematic side elevation of Nayon's Calorimeter. The view has been mounted upside down.

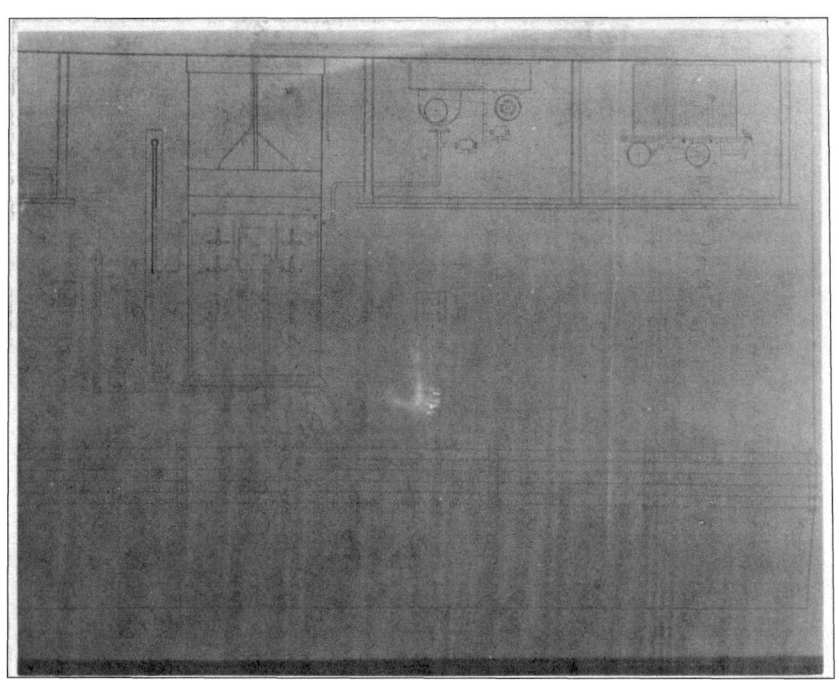

Fig. 66 — Photograph of a chart showing the arrangement for ventilating, and regulating the temperature of Noyons' Calorimeter.

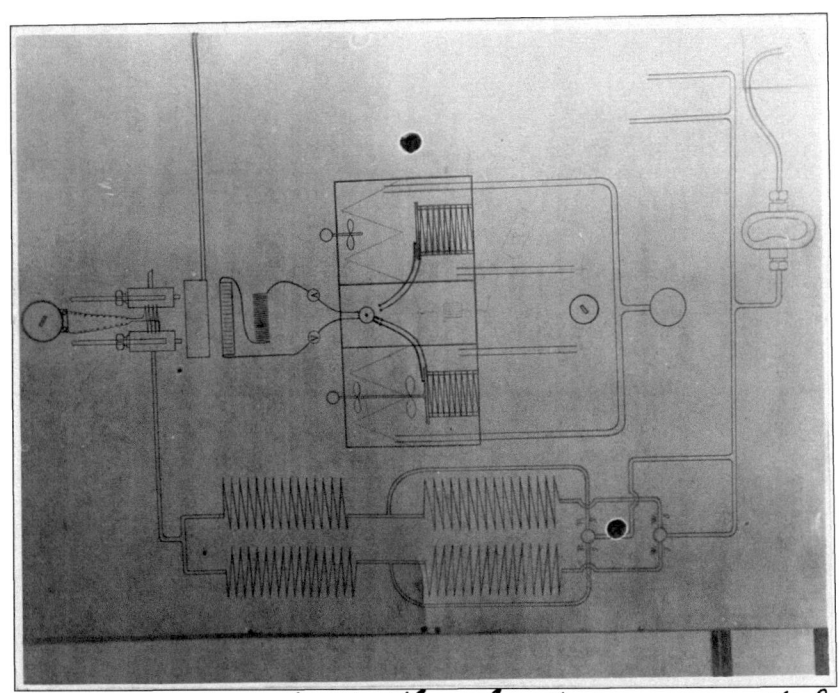

Fig. 65 — Photo of a chart showing the schematic arrangement of the water pipes and electrical system of Noyons' Calorimeter.
View upside down. This page turned wrong way in the binding.

"Rotamesser" (See left hand object in fig. 69) which gives a continuous indication of the numbers of liters of ventilations per minute, a very useful instrument for such work.

The arrangement of the water system of the calorimeter and also of the electrical measuring features is shown in schematic outline in figure 65 and in photographs 67, 68 and 69, something of its nature can be judged. An exact description of all details of these features of Noyons calorimeter must await the publication by himself. He showed me some charts of results which corresponded beautifully with theoretical computation. As a matter of a demonstration we found that the two chambers balanced each other in temperature when they had been unopened and no subject had gone inside. The galvanometer was zero. After placing a subject in one of the chambers and applying electrical heat to the other to compensate the heat given off by the subject we found it possible to secure a balance within a period of five minutes and so be able to compute the heat production of the subject at that time. Noyons thought it would be quite possible to get five minute or even three minute periods with his instrument and by using one of the Cambridge thread recording instruments he plans to be able to get a continuous record of variations during sleep. It seemed to me that Noyons' apparatus was quite practical for attacking a number of problems and I should expect that a considerable amount of first-class work will come from the Louvain Laboratory in the near future, and that this place will be one of special interest to visitors from the Nutrition Laboratory, and for other American workers in the same field. A partial view of their building is seen in figure 70.

Noyons desired to develop techniques which will give him continuous records. For measuring the carbon-dioxide production he plans to use the Ziss Refractometer and to read off continuously the amounts of CO_2 in the expired air from the amount of refraction in these calibrated instruments using one instrument with a standard amount of CO_2 as a comparison basis.

Noyons, with one of his assistants, was working on a electrical method for testing chemical solutions. This involves the use of selenium cells. Two such cells were placed a given distance from a known source of light. In front of these cells square glass vessels such as are used for optical purposes were arranged to be placed with liquid in these glass vessels so that the light would have to pass through the glass vessels to fall on the selinium cell. They then measure the amount of change in the electrical resistance at the cell which is caused by passing the light through a certain solution of some substance and so they measure electrically the quantities of this substance in the solution.

They have much apparatus for working with olfactory and auditory sensation and in fact with all kinds of sensory physiology since this is a field in which Noyons has naturally received much training through his working with Professor Zwaardemaker. Noyons also had developed a measuring arrangement for determining the percentage of two gases which were being mixed. This he expected to use in the study of olfactory sensation and he was interested to learn of the apparatus developed by Hanson. See figure 45, page   .

They had in process of construction a respiration chamber about two meters square made of sheet iron, the top was slanted like the roof of a house. It had been built so that a partition could easily be placed directly through the denter thus reducing the volume of the chamber exactly by half for certain comparison experiments which Noyons said he had in mind.

One of the basement rooms was arranged for an electrocardiographic room. (All of the basement rooms were light, the windows being practically full size. The street in front of the Institute is somewhat higher than the ground level so that one walks in from the street nearly on a level with the first floor. The building sets back a short distance from the street and thus gives plenty of light to the basement windows). While the room has white walls the windows were arranged so that they could be absolutely closed to darken the room. The string galvanometer arrangement is a duplication of that made by Dr. Salomonson of Amsterdam. The room has two large pillars, which can be seen in figure 72 and which are specially built for receiving string galvanometer apparatus some what like the pillars built in the laboratory of Einthoven. The top of the pillar offers space of about one and one half by one meter in area. The arc lamp, a large 20 ampere Seimens & Halske lamp, sat at one end of the pillar. About the string galvanometer a cabinet has been arranged with many windows which slide up and can be taken out when desirable. The glass top is not taken off. This exposes the galvanometer and necessary apparatus and makes it easily accessible during periods of use but keeps it free from dust at other times. (Einthoven has large metal covers which are raised up to the top of the room during the day and lowered covering the whole apparatus table at night. These are opaque). (Salomonson covered his apparatus with heavy paste board boxes made for the purpose). The pillars on which the galvanometer apparatus is mounted are unique in having openings below the top slab (See fig 71) which allow for placing of storage batteries and accessory apparatus as needed. This is very useful and in no way detracts from the necessary solidity of the pillar for the purpose of holding the galvanometer.

The string galvanometer is not just like the large one in Salomonson's laboratory but shows some resemblance. (See figs. 71 and 72 also 76). It is arranged to be water cooled as seen in figure 71.

The camera is the regular large falling-plate design made by Salomonson. It is quite large as can be judged by the photographs but very satisfactory in its operation, in fact, so satisfactory that Professor Einthoven has adopted it for use in his own laboratory, and I do not remember seeing any other kind of camera at that place. A shallow groove around the top edge of the pillar is useful to keep small parts and tools from rolling off the pillar when one is working with the apparatus.

I should encourage anybody intending to plan or build a new physiological laboratory to visit Noyons' and his new Institute. I think that in some ways it is better arranged and more practical than the one built by Professor Hamburger in Groningen and this is saying a great deal. Noyons is undoubtedly a remarkable man in his ability to plan and build. It is to be hoped that he will demonstrate equal originality and versatility in the production of scientific results.

May 14 6 P.M.
Walked out thru the "Backs" to
Bancroft's home for dinner
13 Grange Rd
They are splendid. Mrs. B. was Miss Ball
daut. of the astronomer.
They were interested in Houdini's performance

In garden a wonderful seat all
padded and tucked in.
                    "Don't Waller me"
Waller stories. "Now you must not Waller-me"
an expression in their home.
The Bancrofts' one day at Windsor Castle,
Waller drive them down.
Several of the kind. We greatly enjoy this
secret telling —

I see the boy of 11 asleep.
The other one away.
       Have quite a basin for [franticly?] with the Bancroft
walk back. could not raise Adrian, door locked.

[vertical margin note: First time I rode their place outside]

Cambridge Address
Prof. Sir
Dr. & Mrs. J. Bancroft, 13 Grange Road.
     Cambridge, England. [new addres.]
Dr.   Rivers
   St. John's College. will be there mid Aug and
our winter me to come up for a day.

BELGIUM

Meet Mr. Hartree.

New Paschen type galv described.

Shows heat work on nerve. Most elaborate setup.

Shows the construction of the thermopile, across or along which they lay nerve when stimulated.

This is the work of A.V. Hill. It is calometric in nature.

Hartree is a physicist came to help Hill, who then went to Manchester.

We had tea. Everybody greatly interested in Hill, he had of Patents board for Ariel invention during war.

Addresses Oxford.

Dr. E. Schuster, 110 Banbury Rd. Phone Oxford 682
Dr. Wm McDougall, The Lawn, Banbury Rd. London. Hampstead 1680
Sherrington, Mechanic, Mr. C.J. O'Neill, Physiolgical Dept. Museum, Oxford.
Profssn George Dreyer, Pathology Dept., Univ of Oxford.
Mr. George F. Hanson, " " " also Lincoln Cll.
Dr. C. Gordon Douglas, St. Johns Cll. Oxford
Dr. J.S. Haldane, Luton Rd.
Dr. H.C. Bazett, Dept. of Physiol.
Dr. H.W. Davies, with Haldane.
Mr. H.F. Pierce, Pathol. Dept. Museum, Oxford
Dr. E.F. Adolph, to J.S. Haldane ?

# The Netherlands

The Netherlands Institute for Nutrition.

Dr. Van Leersum.

Dr. Van Leersum was for some years Professor of Pharmacology and the History of Medicine at the University of Leiden. This seems rather a peculiar combination. It came about as a compromise between two factions who had to do with the University. The medical faculty desired a Professor of Pharmacology whereas a man high in national affairs of that government desired that the subject of the History of Medicine should be taught there. They compromised in securing one man to do both and Dr. Van Leersum was appointed to the place. He had exceedingly poor accommodation for the work and instruction in pharmacology. A laboratory was promised but its actual construction was put off from date to date. He worked in the laboratory during the day time and spent his evenings on the History of Medicine giving some courses in this subject, in which he found a very rich and interesting field and he has become a scholar in the subject. It is not surprising that he devoted considerable time to it expecting that later when he secured his equipment and space he could more in Pharmacology.

From both lines of teaching and research, Van Leersum became interested in in the subject of Nutrition seeing especially during the war the practical importance of this subject for a Nation. He gave up hope of immediate improvement in equipment in the University of Leiden for his work and so resigned that position, taking up again the practise of medicine and sought to interest certain influential and wealthy people in the establishment of a National Institute for Nutrition in Holland. It was not difficult to interest the Dutch in this subject. The city of Amsterdam provided a building site of suitable extent on the Amstel River probably five miles out from the city, and certain prominent individuals particularly Mr. J. Wilmink, president of one of the Amsterdam ship companies, undertook the matter of raising an endowment. Van Leersum was sent to North America to visit laboratories, institutes and organizations which have to do with Nu-

This card given to me by van Leersum.

This is a view of Dr. van Leersum's home 28 Vondelstraat, Amsterdam. The house has a front of about 24 ft. and in style is just like the one at the left of it.

Mrs. van Leersum informed me that such a place cost 26,000 g.

In the front part of the house, the basement is kitchen, 1st doctor's office and hall, 2nd living room with balcony, 3rd chambers. The numbers are on the house walls, not on the doors, and easy to see. The beam which extends out at the top is a regular feature of Amsterdam houses, and used to secure rope and pulley when moving and transferring heavy things in and out.

Professor Wertheim - Salomonson lives in a similar house a few doors to the left in this photo.

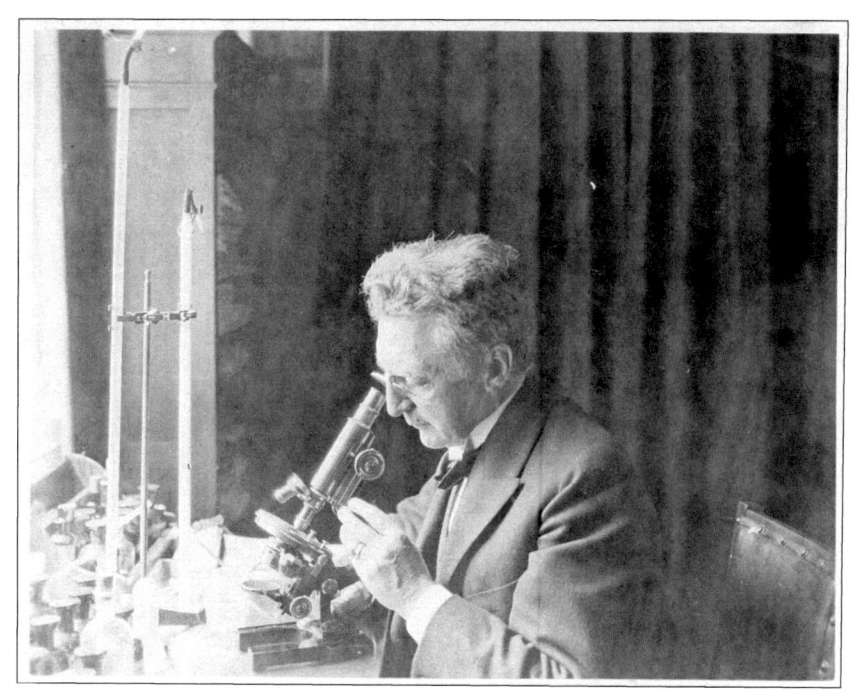

Fig. 73 – Dr. E. C. Van Leersum making an examination in his private laboratory at his residence, 28 Vondelstraat, Amsterdam.

Mrs van Leersum died of heart disease Sept. 9, 1937, Leersum, Holland.

Dr. and Mrs. van Leersum and their son-in-law Mr Wopke de Gavere in the Leersum living room at 28 Vondelstraat, Amsterdam. This is the front room and on the second story, see view of the house on page 162. A small balcony opens off this room by a large door-window and we spent many happy times in this room. Dr. van Leersum's office, two large rooms was on the first floor. Mrs. (Wijsman) van Leersum does wonderful work with her needle and sings at it. Mr. de Gavere, a lawyer, pleasing man. The daughter Jacoba Gavere, had son Ate born May 1-1920, we all visited her and saw the fine boy.

166

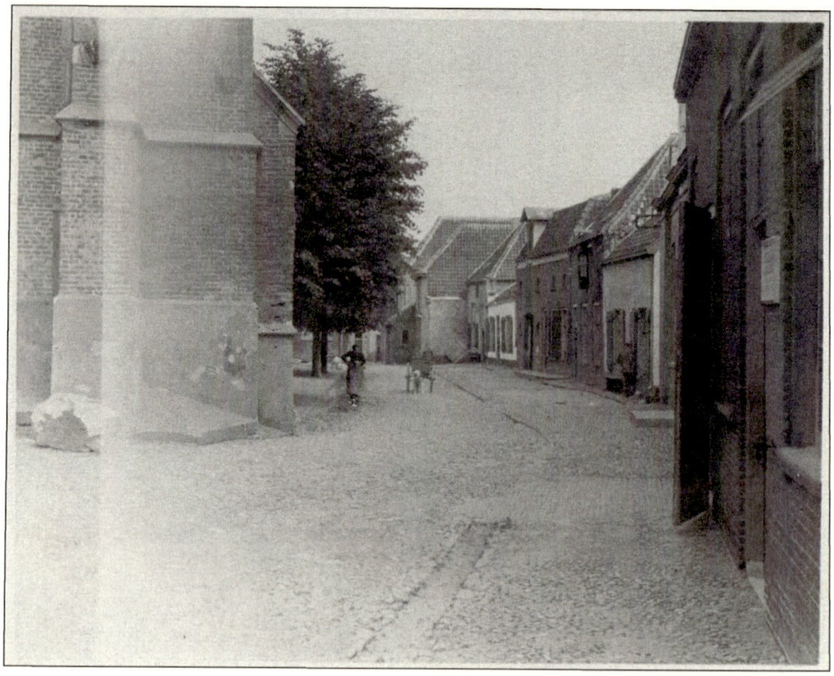

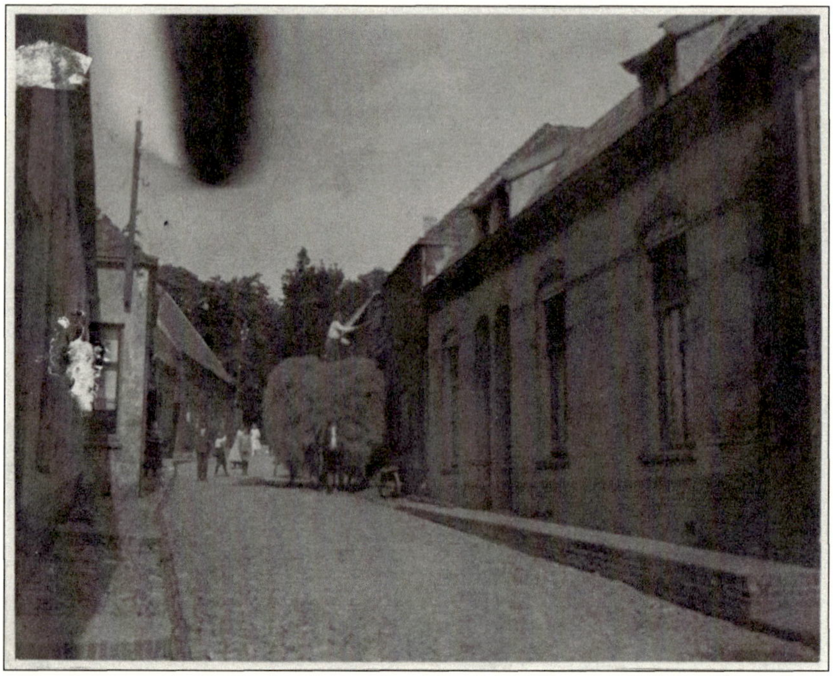

THE NETHERLANDS

trition that he might better decide on the equipment necessary for them and the type of problems that they could best undertake. He was in America from August 27th, 1919 until April 11, 1920 and made a most extended investigation of the nutritional organizations in this country. He was at the Nutrition Laboratory off and on for about two weeks. (For a photograph of Dr. Van Leersum see figure 73.)

It was very pleasant for me to renew my acquaintance with Van Leersum on coming to his home city, Amsterdam, and I cannot here express my great appreciation for his splendid hospitality and many kindnesses during my visit at that place. He generously spent much time in taking me from one laboratory to another in Holland and I thus met several people whom I probably would not otherwise have come to know. He arranged that on Wednesday evening, June 16th, I should give in an auditorium at the University of Amsterdam, a lecture on the equipment and work of the Nutrition Laboratory. This report I supplemented by a large number of lantern slides. Although coming in an examination and vacation period the lecture was attended by a gratifying number and was the first public meeting in connection with the establishment of The Netherlands Institute for Nutrition.

Van Leersum was particularly interested in securing certain apparatus from the Nutrition Laboratory, particularly just now the small respiration apparatus known as the Benedict Portable, an order for which has been duly placed. He plans to build at the very earliest possible, a small house in which to start research and to begin the accumulation of data on certain problems which he feels are especially important to the people of Holland. This small building will finally become the animal building of the Institute after the larger building has been completed. He was in the process of making plans for his Institute at the time of my visit and many of these plans we worked over together in some detail. Van Leersum said that he would certainly visit the Institute at Louvain before commencing to build his own. His attitude and the way he is going at the problem in Holland seems to be in marked contrast to the methid of the French.

It is the plan that The Netherlands Institute for Nutrition will receive some support from their Government but will not be a government bureau in the sense that it has to be supervised by officials and to make regular reports. The Institute in return for support which it receives will supply the Dutch Government with any technical information available on problems which the Government may seek advise.

University of Amsterdam, Neurological Institute and Laboratory.

Dr. Wertheim Salomonson.

Dr. Salomonson is an extraordinary combination of the electrical engineer, the physicist and the physiologist or neurologist. (See fig. 74). He seemed to me remarkably gifted in the ability to bring to perfection such apparatus as the X-Ray equipment, the electrocardiogram and electrical apparatus for treatment and research. Quite recently he had perfected an opthalmoscope with photographic arrangements for photographing the retina of the eye. His photographs are about four cm. in diameter and extremely clear. With these he is easily able to trace the progress of treatment in the cases of retinal tumors and other disturbances.

His string galvanometer equipment is original and extremely fine. He does not mount his galvanometer on a stone pillar but simply on a heavy wooden table. The general nature of the construction of his string galvanometer can be seen from the photographs figures 75 and 76. Also from these something of the arrangement of his compensating resistance box can be seen. He has an ammeter constantly in series with the coils of the galvanometer so that he may at all times know and regulate the field strength of the current which he is using. Through a small accessory telescope mounted between the field coils of the string galvanometer he can view by eye, the string noting its position in reference to the eye pieces of the telescope. Thus he avoids bumping the telescope against the string and so damaging it. (Dr. Bull had a similar arrangement in his own instrument and in the photograph of him, figure 48, he is shown holding this little accessory eye piece). According to the magnetization curves which Salomonson had, his instrument is the most sensitive that has ever been built. He thought it also very practical in reference to the manipulating in mounting the string and in adjusting the

Fig. 74 - Professor J. K. A. Wertheim-Salmonson in his electro-cardiographic laboratory, Neurological Institute, University of Amsterdam.

Fig. 75 – Professor Salmonson's electrocardiographic outfit of his own design.

Fig. 76 – Salmonson's small signal galvanometer at the left. Hill's (Lund) episcotister at the right.

telescopes.

Lately he has perfected a small double string galvanometer which can be used as a signal apparatus in combination with the large string galvanometer and in the same path of light projection. This signal galvanometer is shown in both views at the left of the large galvanometer. Recently, he visited the Cambridge and Paul Scientific Instrument Company in Cambridge, England and arranged for the manufacture of this signal galvanometer by them. (I noticed when I visited that company and signed their guest book that his name came immediately above my own). This small signal galvanometer is electro-magnetic and quite rugged in construction. I think it likely the most practical instrument of this sort which has yet been designed.

At the extreme left, figure 75, may be seen the front of Salomonson's falling-plate camera. A small hood which projects from the face of the camera makes it possible to use it in the room with normal daylight. The camera is not knew. It has been fully described in the literature.

On the right hand side of figure 76 may be seen the episcotister of Hill made in Lund, Sweden. This is the regular electro-magnetic gear drive of the Blix-Sanstrom kymograph. Every gear shaft of the gear arrangement has been extended so that the spoked wheel with its collar may be directly clamped on and so secure the variety of speed. I cannot say just why this was done rather thanunsing the gear shifting arrangement commonly supplied with the kymograph. Salomonson reported that this episcotister when used as supplied by Hill in close proximity with the string galvanometer caused deflections of the galvanometer string due to the breaking and making of the circuit in the electric centrifugal regulator of the kymograph. Hence he has put on a different sort of a centrifugal governor and he made one of a friction tape. He finds that this regulates the speed of the kymograph admirably and his speed may be relied upon. He also is freed from the induced makes and breaks which otherwise occur. For my own part I do not see any advantage in this kind of an episcotister over the usual form which employs a tuning fork to make and break the current to operate a synchronous motor.

In Salomonson's laboratory the strings for the galvanometer are prepared from Wollaston wire in which a platinum core of the desired diameter has been frawn. The original Wollaston wire is cut in the desired length and soldered to the pins which are to be mounted in the string galvanometer. The holder in which the soldering has been done is in the nature of a hinge so that when the wire has been properly soldered to the pins, the holder may be doubled to some extent so that the wire will loop down into a vessel containing the nitric acid for etching off the outer coatings of the wire. Thus the wire is immersed in the nitric acid to points very close to the ends which have been soldered to the two pins. The manipulation is thus very simple and when the etching process is complete, the wire is withdrawn from the acid, the holder is straightened out, and from this holder, the string is mounted directly into the galvanometer. This technique avoids all soldering and extra manipulation after the string has been reduced to the desired size. It is very easy to solder and otherwise manipulate the Wollaston wire and if the attempt at etching is a failure, i. e., if the wire breaks, there is not much loss of time. Salomonson thought it was not necessary to have such a splendid string as those manufactured by Williams and used commonly in his instrument in the United States.

In the case of Salomonson's galvanometer the optical system is adjusted to the string after the string has been centered just as in the case of the galvanometer of Williams. Both the front and back telescopes of the optical system are orientated by operating against two screws and a spring in place of by three s screws as in the Williams instrument. This makes movement much more easily produced than in the case of the latter instrument where one screw must always be loosened before another can be tightened and one has to keep alternating between loosening and tightening of different screws until he gets the telescope where he wants it.

When not is use, Salomonson kept his instruments covered up by heavy paste board boxes which had been especially made for the purpose. This protected them from dust and avoided any part being inadvertently bumped except by the experimenter himself when putting on or taking off the cover. His largest string galvanometer is quite a heavy instrument. He has in operation two complete out fits. The second one which employs a smaller galvanometer can be partly seen in the photograph behind Salomonson, figure 74.

Salomonson showed me a large amount of apparatus which he had designed and had in use for measuring the sensitivity of nerve and muscles to electric shock. He could preform these measurements by three methods; induction shock; capacity discharges or direct current stimulation, and he was interested mathematically to develop a formula which would harmonize the results obtained with all three methods. I sinderely regretted my inability to follow his splendid mathematical reasoning on this problem. He has a research student who is writing his thesis on this subject and Salomonson will himself probably later get out a book on it. Although I was with Salomonson for parts of two days, I felt that I had had only the merest glimpse of him and of his many excellent instruments and techniques.

174

Fig. 77- Professor F. J. J. Buytendyk (standing) and his assistant Dr. M. N. J. Dirken on their rowing-machine ergometer.

Fig. 78- A calorimeter of the Atwater-Benedict type lately constructed by Professor Buytendyk, Amsterdam.

Figs- 79, 80, 81, 82, and 83 in the Nutrition Laboratory copy of this report are plates which were kindly given me by Professor Buytendyk.

THE NETHERLANDS

A photo of Buytendyk's calorimeter given to me by him.

Image Not Reproduced

Image Not Reproduced

Free University, Department of Biology.
Professor Buytendyk and Dr. Dirken.

The Free University in Amsterdam was established by a religious organization and I have been informed that due to theological tenets of this organization, the theory of evolution and the practise of vivisection are tabooed in the University, hence Professor Buytendyk has found it best to study such problems as the learning of animals, respiration physiology, and work physiology. He seemed unduly appreciative of my visit (I was accompanied by Dr. Van Leersum) and I was very glad to see something of his work in which he is ably assisted by Dr. Dirken. They publish most of their results in English in the "Koninklyke Akademic Van Wetenschappen Te Amsterdam". These proceedings, I think, we should arrange to receive at Nutrition Laboratory, because many of the Dutch publish here.

A piece of research work which I understood was nearly finished was that on the subject of muscular work. They measured the CO_2 production, the changing of pulse rate, etc., etc. and used for the work a rowing machine adapted as an ergometer (See fig. 77). A paper recording mechanism has been attached which can be seen in the lower right corner of the photograph and a graphic record of the work is thus obtained. Dr. Dirken is sitting on the ergometer in position for work, Professor Buytendyk is standing by his side. It seems to me their ergometer has some very good points. It offers work which involves all the main muscles of the organism and the work is periodic in nature.

Their building is new, designed for the purpose, quite adequate to their needs, and it seems to me quite convenient. While reasonably well finished, it shows clearly that money had not been wasted. The very top floor they use for animals particularly in studying learning experiments. They had one monkey on which they had been doing some work. On the middle floor was a respiration laboratory for work with small animals and microscopic organisms, also a laboratory for studying muscular work. My recollection of the first floor is that it was classrooms and offices, and in the basement, they had in process of construction, a respiration calorimeter built along the lines of the Benedict-Atwater apparatus. (See fig. 78.) This photograph shows the chamber, the scales and the water collecting arrangements and to some extent, the control table. The construction was almost wholly without the use of iron and had been done very cheaply. I do not know whether they will be able to get satisfactory results with the apparatus as they now have it. I did not go into the details of its construction with Buytendyk and he did not show me data in the nature of alchol checks.

Behind the calorimeter, they have one of the Holima Autofrigor equipments installed. (See catalogue in Nutrition Laboratory office.) This apparatus they found very satisfactory for controlling the temperature. It interested Dr. Van Leersum as a prospective equipment for his Institute.

Near the calorimeter and under the stairs by which the basement is entered they had a chamber approximately one meter in diameter and two and one half meters tall, circular in shape which was arranged to be entered from the top. This chamber could be filled with water the temperature of which was controled by their refrigeration machine and they contemplated making certain respiration measurement with the subject completely immersed in water wearing a gas mask. They also had work in mind in reference to the volume of the body determined by this immersion chamber)

On one of the tables in the laboratory where I found the ergometer I saw about 40 small William's acid bottles. Buytendyk said that these had been made by the V.F.L. for some one in the United States he thought the Nutrition Laboratory but since the V.F.L. could not deliver them, he has been able to buy them recently at about 20 cents a piece. I did not know of our having an order placed for this apparatus but it illustrates how the people in the small neutral nations near Germany were able at the end of the war to rush in and buy apparatus for a very small figure, while the exchange of German money was favorable to them and the German prices had not greatly increased.

Buytendyk showed me a micro-gas analysis apparatus, after Krogh, figure 81 to which he had added a small chamber. In this chamber he had micro-organisms and so collected the CO_2 production from them. He had arranged a new gas analysis apparatus not yet described figures 80, 82 and 83. In this apparatus his changes were in _pressure_ rather than in _volume._ The gas sample to be analysed and the potash are put into two adjoining compartments (See fig. 82) which may be connected when desired. The other side of the gas sample is connected to one side of a small manometer with a drop of oil in it. (See fig. 83.) The gas is admitted to the potash and the CO_2 absorbed. The pressure is changed in the compensating chamber by a mercury column until the manometer is again central. Then the column of mercury is read for the change in pressure caused by the absorption of the CO_2. He says that the mercury is always dry and clean. He would like very much to have some one else try this apparatus before describing it. A larger apparatus of this nature is shown in figure 80. He had arranged a long mercury column using a beaded shaped tube on one side for rough adjustment, and a cappillary column on the other side for fine adjustment. The sections between the bulbs on the rough side are of cappillary size. At the bottom of the column is a glass piston moved in and out to adjust the the height of the mercury. He seems to welcome criticism of which I was unfortunately not able able to supply much in reference to gas analysis apparatus.

Buytendyk was well pleased with an Aquarium which they had made of cement and glass and in which they kept sea-organisms in salt water. (See fig. 79.)

I hope that if Dr. Benedict or Dr. Carpenter are again in Amsterdam they will take the opportunity to call on Buytendyk and see his apparatus. I think probably he should receive some of the publications from the Nutrition Laboratory.

After visiting Buytendyk, one of his students, Dr. Van der Hyde, who had been working with him in animal psychology called on me at the residence of Dr. Van Leersum to talk over the prospects of an offer to go the West Virginia University at Morgantown, West Virginia to the department of Professor Morse, in Psychology. He was somewhat surprised that I could not give him more details and information about this University and its environment.

State University of Utrecht,
Institute of Physiology and Physiolological Chemistry.
Professors Pekelharing, Zwaardemaker and Ringer.

On my visit to Utrecht I had the pleasure of being accompanied by Dr. Van Leersum. We called at the home of Professor Pekelharing, who has now retired from actual service, to give him our compliments and also greetings from the Nutrition Laboratory. We found Pekelharing in his library, he had been working at his typewriter on which he does all his writing. (See figs. 84 and 85.) He did not seem at all feeble but was animated and engaging in his conversation although he complained of being old and put on the shelf. He said he found much of interest to do in writing and in conferring with colleagues. Professor Ringer came to see him about some administrative matter while we were there. Pekelharing was much interested in hearing about the Nutrition Laboratory and its workers and also questioned Dr. Van Leersum concerning his trip to America. Pekelharing is a splendid character and it was something of a benediction to meet him.

In Professor Zwaardemaker, I was quite surprised to find such a husky, quick and young appearing man for I am certain he must be sixty years or older. (See fig. 86.) Throughout his laboratory I saw much of both psychological and physiological interest. In his laboratory for studying olfactory sensation, he showed me the latest forms of his olfactometer for measuring the physical stimuli for olfactory sensation and for properly grading the stimuli. He reported that he could entirely deodorize a tube or vessel by subjecting it to ultra-violet light, but said that, of course, the nose itself could never be deodorized, for there is always present in the nose under the very best circumstances, the smell of the individual's blood. He gave a very beautiful demonstration of the static electric charge which may be produced by forcing certain liquids through an atomizer by compressed air. It seems that a substance that has no odor such as water when forced through the atomizer causes no static charge while any substance that has an odor will cause a static

Fig. 86 – Professor H. Zwaardemaker at his desk, University of Utrecht.

Fig. 87 – In Professor Zwaardemaker's office.
Drs. Gryns, Zwaardemaker, and van Leersum.

Fig. 84 – Professor C. A. Pekelharing at his typewriter in his residence library, Utrecht. Dr. van Leersum at left.

Fig. 85 – Dr. van Leersum and Professor Pekelharing.

charge to be taken up by the gold leaf indicator.

In a room devoted to research on the analysis of sound waves, particularly the sound of the voice, Zwaardemaker had a great number of resonators for collecting certain partials of the sound complex. These he had connected in such a way that he could register the intensity of the tones selected by each. He also demonstrated to me a very simple and objective method of securing a graphic sound wave. Taking a long length of glass tubing about a meter and one-half long and five cm. inside diameter, a teaspoonful of very light powder is put in the upper end of the tube and the tube is placed on an incline. Then the subject sings the tone quite loudly in the upper end of the tube next the powder and the powder rolls down forming itself into waves corresponding to the nodes and crests of the sound wave which is being produced. Zwaardemaker, as everybody knows, has a very remarkable equipment of instruments with a dark room and a soundproof room. It would take much more than one afternoon to properly visit his laboratory. He told me something about his recent work with radium and radio-activity in relation to the organism and particularly its action on the heart. He has a series of researches in reference to radio-active equilibria and the influence of uranium and also of potassium which opens up many new problems in physiology. At Louvain, they were also working on the effects of potassium and calcium on certain nerve endings, for example, in the iris of the frog's eye. While with Zwaardemaker, Dr. Gryns, who is interested in scientific agriculture and plans to visit the United States in the fall of 1920, came to confer with Dr. Van Leersum and myself about arranging his trip. I persuaded the three to sit for the camera, see fig. 87. One wishes also that we in America might have the favor of a visit from Zwaardemaker.

In the old laboratory of Donders now presided over by Dr. Ringer since the resignation of Pekelharing, I saw the equipment for teaching and research in physiological chemistry. No new apparatus was called to my attention, Dr. Ringer having recently come into this position has been busying himself with matters of administation.

State University of Utrecht, Institute of Inorganic Chemistry.
Dr. Cohen.

The best known and most prominent student of J. van't Hoff is Ernst Cohen, who in the Van't Hoff laboratory at the University of Utrecht is working in a department of chemistry, connected with that of his teacher; namely, Allotrop, the influence of pressure on chemical equilibrium. Cohen who is a close friend to Dr. Van Leersum very kindly showed me his wonderful arrangements for producing high pressure on chemical substances and for maintaining these pressures uniformly for long periods. I noticed in his laboratory a number of stirring baths all of which were operated by small heat engines in place of using an electric motor. He explained that these engines gave a slower and much more satisfactory movement of the liquid than is obtained with motors unless gear reductions are employed which are clumsy. He also pointed out that the engines worked with a very small gas flame and are quite economical in their operation. I saw them in several of the Dutch Laboratories but especially in the laboratory of Cohen.

I was thoroughly interested in the shop of Cohen's laboratory. They had one or two young mechanicians who worked in the shop all the time and a head mechanician who combined the duties of draftsman, photographer and mechanician. This man makes drawings of apparatus to be constructed, does delicate parts of the construction himself, photographs the apparatus and prepares illustrations of it for publication, makes the lantern slides, prepares the charts and diagraming of data for publication, photographs these charts for sending to the printer, etc. With all due appreciation to the genius of Cohen, I can see that this man is a most useful individual in that laboratory. He finds it best in preparing illustrations for the printer to draw his figures fairly large and then photograph them on a plate about $6\frac{1}{2}$ x $6\frac{1}{2}$ and send the photographic print with the manuscript for reproduction. The engraver can make his plate from the photographic print as easily as from the large original drawing and and it is much less clumsy to send and to handle, and, furthermore, the original drawing runs no chance of being spoiled or destroyed.

State University of Leiden, Physiological Laboratory.

Professor Einthoven.

For one who is interested in the string galvanometer and its uses, it is particularly attractive to visit Einthoven, the originator of this instrument, and to see the splendid laboratory which he has developed especially for work with the string galvanometer. In figure 68, Einthoven is shown standing by his original galvanometer in his special laboratory. Scientifically speaking, I think Einthoven is entirely engrossed in the theoretical problems associated with the string galvanometer. Having originated it, he naturally feels, being a man who is extremely thorough, that he wants to work out the several problems connected with it in a way which will make it unnecessary to ever repeat the work. Hence he has built his laboratory with a special view to having it free from vibration, making extra large brick pillars on which to mount the instruments. He has an overhead traveling crane for the moving of heavy instruments without the danger of touching or breaking delicate parts just as one would have in a power plant for shifting dynamos. There are also conduits in the floor to accommodate wiring and to facilitate the changing of the same as experimental needs would demand. There are big spaces for the shifting of cameras and registering devices. The walls are white and the illumination is entirely from a skylight. This skylight, when desirable, may be closed off by a curtain operated by motor, like the large curtains in the Harvard Medical School Auditoriums, i. e., in the auditorium for physiology and bio-chemistry. The equipment seemed very complete.

At the time of my visit, I was in some ways fortunate and in other ways not. It was a pleasure and profitable to meet Professor Verzar of Budapest who was previously an assistant in the laboratory of Professor Tangl but was now working in Holland since he found it quite impossible to do anything in

Fig. 88 – Professor W. Einthoven photographed by his original string galvanometer in his special laboratory for string galvanometer work, Leiden.

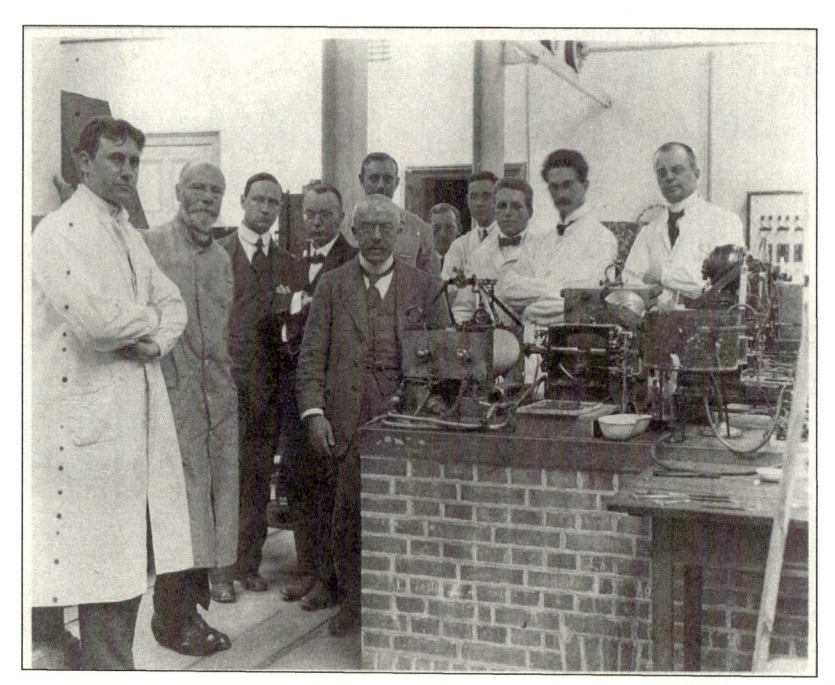

Fig. 89 – Group of scientific workers in Professor Einthoven's laboratory. Bytel, Einthoven, Prof. Verzar (Budapest), Dr. Liljestrand (Stockholm), Bergansius.

Please mark names on this and return to W. Rm

1. Dr. de Jong
2. F.W.N. Hugenholtz assistant
3. v. d. Bijl
4. v. Lawick van Pabst assistant
5. Hoogewerf
6. van Hinte mechanician
7. Dr. Bergansius conservator
8. Dr. G. Liljestrand, Stockholm
9. Dr. F. Vertzar, Hungary
10. Professor Willem Einthoven
11. Dr. F. Bijtel

This brick pillar is one 2, 3 & 4 in this large room.

Fig. 90 – Second view of the group with Professor Einthoven June 18, 1920. A better likeness of Einthoven.

THE NETHERLANDS

185

(Area Intentionally Blank)

Physiological Laboratory of Professor Einthoven
Leyden.
1920

Image Not Reproduced

Austria. Also I was very glad to meet Dr. Liljestrand of Stockholm, a man who with Dr. Krogh and Stenstrom, has worked in respiration physiology and at present is completing a course in pharmacology with Professor Magnus at the University of Utrecht. Both of these gentlemen had been desiring an opportunity of visiting the laboratory of Einthoven, hence, when Dr. Van Leeuwen learned that I was to visit Einthoven he arranged for these other gentlemen to be there at the same time. Einthoven's colleague, Dr.   a physicist, and all his assistants left off their work and devoted themselves to receiving us three visitors. We were, therefore, shown the instruments and equipment in general but there was no good opportunity for me to ask questions about the details and to learn as much as I could have done if alone with Einthoven. After we had looked at the instruments, particularly at the original string galvanometer which is shown on the corner of the pillar in all the views, figures 88, 89 and 90, and had seen in some detail an instrument which when used with the string galvanometer, registers the amount of vibration of the pillar and the building, it was suggested that an electrocardiogram be taken of the three visitors as a memento of the occasion. During this while, which was indeed very thoughtful of them, Einthoven talked about the German Physiological Meeting and the Paris Meeting which was to come in July and of the possibility of getting back to the old basis of truly International Scientific Meetings, also in reference to the possibility of the Physiological Congress coming to the United States. Following the taking of the electrocardiogram I had an opportunity to take the photograph shown herewith.

Einthoven is an humble and extremely retiring man. I afterwards saw much of him at the Physiological Meeting at Paris and found him most friendly and companionable. It seemed to me that more than in the case of the other older and prominent men, Einthoven was usually talking in any crowd with some of the younger men.. Although in sympathy with the Allies during the war, according to statements which were made to me in Leiden, I think Einthoven felt perhaps unusually reticent at the Physiological Congress in Paris. I am sorry that circumstances were such that I did not learn just what research work was in progress in his laboratory at that time.

Dr. Liljestrand felt that in studying Pharmacology, he was putting himself in a better position for election to some chair that if he continued only in respiration physiology. Although a doctor of medicine, he prefers research work to medical practise. Dr. Verzar, whom Dr. Benedict will remember having met in Tangl's laboratory, is a very bright man and full of research ability. He was quite discouraged about conditions in Austria and did not know when it would be possible for him to return there and to assume his teaching duties as professor of experimental pathology in the University of Dezrecescu, Hungary. In the meantime, he was working in the pharmacological laboratory in Leiden with Professor Van Leeuwen.

State University of Leiden, Pharmaco-Therapeutic Institute.
Professor Van Leeuwen.

Professor Van Leeuwen had visited the Nutrition Laboratory within the year, having been brought here one afternoon by Dr. Reid Hunt. He was with us probably two hours, an hour and one-half of which was spent with Dr. Benedict and a half hour with myself. He had prepared a report in book form of his visit to America which has been confined to the Medical centers of Baltimore, New York and Boston, spending by far the smaller amount of time in Boston. In the report he describes with a good deal of detail his visit to the Nutrition Laboratory and the inspection of our apparatus. He has been elected to succeed Dr. Van Leersum as Professor of Pharmacology at the University of Leiden, but does not have combined with this the work in the History of Medicine, that subject having been dropped at the University. (I think it is an important subject full of live interest and should not be dropped when once having been given a good start particularly at this old school.) I may say parenthetically that the old building of the University of Leiden is one of the most interesting to visit. Its hallways and faculty room are uniquely decorated. Van Leeuwen still has no better quarters for his department than at the time Van Leersum was teaching pharmacology. However, he has a promise of the old library building for use during the next two or three years during which time he is to plan and superintend the construction of a new laboratory.

Van Leeuwen had recently been interested in constructing a chamber for dogs and combining with it Dr. Benedict's arrangement for registering the activity of the animals. One side of the chamber rested on the two points while the other was suspended by a pneumograph. Van Leeuwen injected certain drugs into the animals and by this apparatus registered changes in the general activity of the animal after such injection. He had just written for publication a description of the apparatus in English (he gave Dr. Benedict full credit) and he besought me to edit this article

Fig. 91 – Professor W. Storm van Leeuwen at his special circular desk at his residence.

which I did with pleasure although his use of the English language is entirely satisfactory for the scientific reader.

I found Van Leeuwen an exceptionally keen, active and sympathetic man. (See fig. 91.) He complained that after visiting America he was afflicted for some time with what he called "the American sickness" by which he meant the spirit of depression after having seen "so many fine laboratories with such splendid buildings". I think, however, that the American upon visiting European laboratories is equally staggered at the immensity of the work which has been accomplished by those scientific men across the water. Van Leeuwen attended the German Physiological meeting in Hamburg and told me considerable about that occasion. He had recently gotten together a small amount of money and subscribed to some of the American physiological and bio-chemical journals, giving this subscriptions to a group of German scientists in the form of a magazine club, each man in this group having the journals for about five days in the month. Professor Kestner at Hamburg is one of those who receives these journals and he spoke to me very feelingly of his appreciation for the kindness on the part of Van Leeuwen.

State University of Leiden, Physical Laboratory.
Drs. Onnes and Crommelin.

In company with Dr. Van Leeuwen, I visited the Physical Laboratory of Leiden which had come to be known as the Cryogenic Laboratory due to their specialization in regard to the physics of low temperature work. This laboratory has been largely developed and is at present directed by the famous Dr. H. Kamerlingh Onnes who, at the time of our visit, was absent. We were fortunate in meeting Dr. Crommelin, the associate of Dr. Onnes, and he kindly showed us through their laboratory. It is, as everybody knows, a very specialized place and they employ a large amount of machinery in the several liquefying equipments which they keep continuously in operation. The method is, I believe, called "the cascade process". First, they liquefy methylchloride, then with this they are able to liquefy ethylene and step by step proceed with oxygen, air and hydrogen successively. Professor Onnes, in 1908, succeeded in liquefying helium in his laboratory.

Naturally this laboratory is interested in the making of thermometers and particularly in glass vacuum flasks for receiving temporarily the liquefied substances. So it is perhaps a natural development that they have connected with this laboratory a school for instrument makers. Of this school, Dr. Onnes is director and Dr. Crommelin secretary. They have a very thorough course in wood-work, glass blowing, and machining. The boys receive a thorough training in these branches and some of the things which they produce are sold. They have an equipment in the practice shop of I should say about fifteen foot power lathes and as many small benches. Most of the mechanics for the laboratories in Holland come from this school, which supplies also places outside of Holland. Dr. Crommelin gave me the annual reports of this school for the last three years. They have a membership list which contains the names of some well known scientific men in America. Information about the school may at some time be useful to us at this laboratory.

Crommelin told Van Leeuwen and myself something about the recent work on what they term the supra conductive state in metals which is brought about by subjecting them to the temperature of liquefied helium. In this condition the electrical resistance of the metals is greatly reduced so that electric current can circulate in them for a long period. At such visits one devoutly wishes that he knew more so he could understand and ask more but the days of Aristotelian knowledge have passed for most men.

State University of Groningen, Physiological Laboratory.
Professor Hamburger.

Professor Hamburger had been ill from an infection but when I visited there was practically well except for experiencing some weakness on physical exertion. Figure 92 shows a photograph of him standing at his laboratory desk. He accompanied me to the Physiological Institute, a general view of which is shown in figure 93 and with his capable assistants, Brinkman, Ferings, Weinberg and Creveld, showed me completely that excellent Institute. I was especially interested in their work in clinical physiology which corresponds to some extent to the course which is given by Dr. Bazett in Oxford. They used the Wiersma method of recording the pulse rate from a small especially shaped rubber bulb strapped in the hand. I was interested in securing one or two of these rubber bulbs which they assured me could be procured from the manufacturer, Boulitte, in Paris. (When I visited Boulitte at their shop and also at their exhibit of apparatus during the Physiological Congress I was unable to secure this bulb and they did not seem to know anything about it). It is unnecessary for me to go into any great detail about the equipment of this laboratory in Groningen. Dr. Benedict has for me very satisfactorily characterized its equipment in the report of his Foreign Visit to Laboratories in 1913. That description is absolutely correct for 1920, in fact, it rather astonishes me. One had the impression in going into the laboratory that there was nothing especially in progress. Everything seemed definitely in its place either in a storage closet with glass doors or satisfactorily covered up. The card indexes for equipment in each room were all intact as well as the reprint file in which reprints were placed in order as received. This file was just about full to capacity and Hamburger was going to have to put in another cabinet to receive future reprints.

In the basement they had rooms expecially adapted for temperature experiments both for refrigeration and for torid temperatures. I think these had been lately equipped. Apparently they had spent a large amount of time in prefecting these arrangements. I thought the place seemed quite undermaned, I remember that there was ohly one mechanisian. The machine shop like all the rest of the place was a marvelously fine shop. They had on hand quite a lot of stock which was verynnicely assorted and arranged. I remember especially their great stock of screws of all sizes and quite large quantitées of each. They have several acres of farm land connected with the laboratory particularly adapting the laboratory for experiments with domestic animals, cattle and horses. Hamburger, before coming to the University of Groningen, was interested in veterinary physiology, consequently this is the reason for the fine laboratory for large animals which is the left hand wing shown in the general view of the Institute, figure 93, with the very elaborate room for operations on horses. I had the impression that the large number of arc lamps in use in this room referred to on page 311 of Dr. Benedict's report were not for motion pictures but for the purpose of illumination while operating on horses and cattle. It was in this room that Professor Starling chose to perform his demonstration operations at the time of the Congress in 1913.

I was sorry not to be able to take a photograph of Hamburger in his elegant office. The hangings were so large and luxurious and the light so soft that it was impossible, particularly as he has quite a marked tremor. The only place I saw which was satisfactory for a picture of him at the Institute was in his private laboratory standing by his laboratory desk. Even this picture which was a rather short exposure, shows a slight amount of movement on his part. The photo taken out in his garden is quite satisfactory,(see fig. 94).

There was a good laboratory in the basement for cardiographic work and for electro-physiology. It was well equipped with modern apparatus but no researches were in progress.

Fig. 92 – Professor H. J. Hamburger standing at his laboratory desk, Groningen.

Fig. 94 – Professor and Mrs. Hamburger in their garden.
Praediniussingel 2, Groningen, Holland.

Fig. 95 – A memorial in tile given Professor Hamburger by citizens of Groningen and placed in the Physiological Institute after the IX International Physiological Congress.

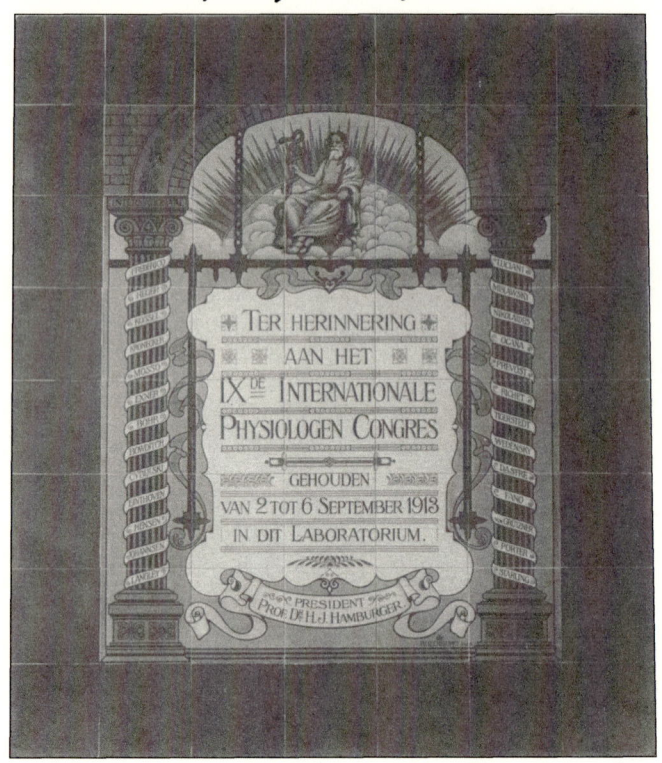

Hamburger, undoubtedly, delights in organization and has marked ability for it. It was these features of his laboratory that he seemed to take particular pride in showing to me. Each laboratory room according to his scheme is more or less independent and complete in itself, having a card index and a definite arrangement for all chemicals and for each piece of apparatus. Dr. Benedict, in his 1913 report, page 311, speaks of the practical idea of a small tool kit for each room. To a certain extent I think these outfits are useful in bringing the tools to the work but many times it is more convenient to take the work to the shop and I thought that there were more of these kits sitting about than could probably be useful. Another item which was in each room and should not be forgotten was a box of sand with a small shovel in it. At first, I thought this was a spittoon arrangement but Hamburger pointed out that it was a matter of precaution against fire. These boxes sit on the floor and are at best, I think, difficult to keep looking tidy. They told me it had never been necessary to use any of this sand to extinguish a fire since the laboratory was equipped seven or eight years ago. The walls of the hallways particularly on the second floor and along the stairway are covered with pictures of physiologists and of physiological subjects and photographs of historical physiological prints, etc., etc. These are of a uniform size and have been photographed and gathered by Hamburger or under his direction. Making them of uniform size simplifies the problem of framing. The pictures are numbered and reference is made to them in the lectures in physiology hence they are made useful in the teaching, I think probably, to good advantage. There is likely no one who has more ability or interest than Hamburger has in arranging such details. The care which he demonstrated in arranging the details for the Physiological Congress held at Groningen in 1913 is almost unique among scientific workers. I imagine that that Congress took more out of Hamburger than two or three years at ordinary work. It was at his behest that the street car company of Groningen laid a special extension of their line out to the Institute. This saved the members of the Congress many steps in walking.

(The extension is now called the Hamburger Line). As an example of his insight in helping and preparing for strangers, I might mention that when there, he invited some members of the Groningen faculty to his home that I might have the pleasure of meeting them. In anticipation of any difficulty that he thought I might experience in getting the names of the men who were to be present and understanding something as to their interests, he without any request on my own part, provided me with a paper on which he had written out plainly the names of the men, their status in the faculty and the departments in which they were professors. This was a great help in giving one a little preliminary familiarity with the names and it is, I think, a truly characteristic thing showing the insight and ability for attending to details which makes Hamburger so well loved. After the meeting of the Physiological Congress in 1913, the citizens of Groningen presented Hamburger with a memorial in tile, a photograph of which is shown in figure 95. This memorial is placed in the wall of the entrance hallway at the Physiological Institute.

One of the very pleasant experiences which I had in Groningen through the kindness of Hamburger was the meeting with Professor J. C. Kapteyn, the eminent astronomer, whose "plan of selected areas" drawn up in 1906 as a basis for a thorough investigation of the constitution of the stellar universe has provided a program in which the astronomical observatories of all nations are cooperating. This genius himself has no observatory. His department now occupies what was previously the old Physiological laboratory. When going over to visit Heymans, I saw Kapteyn working in his shirt sleeves at his desk. As a research associate of the Solar Observatory on Mt. Wilson of the Carnegie Institution of Washington, he has come into close touch with American astronomers. He has been in the United States a great many times and thinks very highly of this country. The topics of conversation ranged all the way from gum chewing in the United States to the latest development of the "star-stream theory." Among other statements he assured me of his belief that women were more satisfactory than men as a usual thing as laboratory assistants. One never meets a more companionable sort of a scientific man than is Kapteyn. He is so free, open-minded and liberal and withal so keen and has such a pleasant way that I can easily understand how all scientific men are willing to cooperate with him in the great project of systematically collecting astronomical data.

State University of Groningen, Psychological Laboratory.

Professor Heymans and Dr. Brugmans.

In company with Professor Hamburger, I visited the Psychological Laboratory which is in the main University building. This building is quite a grand structure and I was impressed that the psychological rooms looked like they might be absolutely new rather then having been in use for some eight or teh years. Professor Heymans has a very large dark and sound proof room. The ceiling of this seemed to me remarkably high for such a room. Wires were stretched across above something like a wireless aerial but this was for taking voltages of desired strength for use with apparatus in that room. As usual, they found difficulty in kaking the room entirely sound proof and had combined certain hangings around the walls to help deaden the sound. I believe they were working more on problems of vision than on sound. They had some nice apparatus of the usual types found in psychological laboratories and Heymans had developed some especially fine steropticonsviewsand teaching apparatus of this character. He had stereoscopes made with a red glass for one eye and a g green glass for the other. These were of extremely simple form for class demonstration and much like a common lorgnette. Then he had developed a number of complicated lantern slides for use in projection work. These lantern slides allowed of the movement of figures which were presented. This was especially for studying illusions of movement also for presenting the effects of the Muller-Lyer illusion. Heymans and Brugmans had lately been cooperating on a study in reference to the required to read a group of words, names of colors, arranged in a square as compared with the time required to call off the names of colors when the colors were mounted on the chart in place of the words in the exact location that the work had previously had. They find it possible to read off the words considerable quicker than they can call off the colors. In all cases it involves the same amount of speaking. I suggested the difference might possibly be due to more indefinite succession of visual fixation and more time used in general fixation in the case of the colors. One should take eye movements records during the experiment to prove the point. It would be of interest.

They had in progress at this laboratory an experiment on thought transference. The arrangements were something as follows: In a basement room which was quiet and not very light was placed a table with a screen about halfway along the top of the table. The subject, a student, who, according to report had made the discovery that he could read the thoughts of others, sat behing the screen and put his right arm under the cloth screen so that his hand could rest on a board which was arranged something like a checker-board. The squares of this board were numbered and each square was about 2x2 inches. Directly above the part of the table on which the checker-board was placed, a window had been arranged in the floor of the upper room which was their main laboratory room. An observer, Heymans or Brugmans, sat immediately above the subject and looking down through this window, watched the subject's hand move on the checker-board, and in this way the subject could not possibly see the experimenter or visa versa. All destracting noises seemed to be avoided, there was nothing heard of the breathing of the experimenter of any slight movement on his part which would give secondary criteria to be judged by the subject. The method then was that the experimenter selected according to a chance order which he had arranged, one of the numbered squares of the checkerboard. He enters this number in a table and then gave a signal by tapping on the glass with a stick that the subject was to place his finger on the number which had been selected Meanwhile, the experimenter thought of this number and looking down through the window, kept his vision fixed on the square, where he wanted the subject finally to place his finger. The subject moved his hand slowly over the checkerboard and finally brought his finger to rest on some square. He kept it here until the next signal to begin hunting. The experimenter recorded the

the number of the square pointed to by the subject and so the experiment proceeded. According to what Brugmans told me and the results which he showed me, they had many more correct answers than could be accounted for by the theory of chance coincidence and we were engaged in discussing for a long while the possible cues which the subject could gain from the experimenter and by which he could regulate his choice. Brugmans agreed that they must arrange some signal apparatus which was much less subjective than tapping on the window with a stock and also that the subject should have some way of making a definite and uniform signal when he was satisfied with his position. The arrangement and the experiment were occupying Brugmans who appealed to me as a very earnest and unbiased investigator, ready to do anything and everything to make the experiment free from any known or supposed secondary cues for the subject. Brugmans exhibited great interest in the physiological and psychological apparatus of the Nutrition Laboratory as shown in the photographs which I had with me, and I am glad to have established contact with the laboratory.

Denmark

University of Copenhagen, Laboratory for Animal Physiology.
Professor Krogh.

I found Professor Krogh and Dr. Marie Krogh, his wife, busily engaged in the laboratory. Their immediate task was reading the proof of a large article shortly to appear in the Journal of Bio-chemistry on the subject of "The Relative Efficiency for Muscular Work of Carbohydrate and Fat Diets." This research was done in collaboration with three others and impressed me as a very fine piece of work. I went over a good deal of the proof. They found that carbohydrate diet was more efficient for muscular work by an amount on the average of about 10%. Figure 97 is a photograph showing the respiration chamber which they employed in this research together with bicycle ergometer. The ergometer was placed, of course, in the chamber which upon the subject's entering is tilted up as shown in the photograph. The upper movable part fits into a water seal during the experiment. The chamber is fitted with a number of windows and seemed of suitable size for bicycle work. Krogh strongly favors the method of using a chamber and supplementing this with gas analysis. He said that <u>invariably</u> in his own work, the use of a mouth piece or a mask disturbs the result making the respiration somewhat abnormal.

Krogh hopes to continue the research on muscular work, next time investigating the efficiency of alcohol as food for work. He expects to use the technique of Widmarks of Lund to determine the amounts of alcohol in blood and urine. He referred to several papers of Widmark and spoke highly of him, urging me to visit him on the way to Stockholm. Krogh thinks that by feeding such foods as to keep the respiratory quotient high and then giving alcohol, he can tell more what happened than if the respiratory quotient is low. But he is very much afraid that small traces of the alcohol in the gases will destroy the accuracy of the gas-analysis. He has made no preliminary experiments along this line

Fig. 96 - Professor August Krogh in his laboratory studying the capillary circulation of blood, Copenhagen.

Fig. 97 - Respiration chamber and bicycle ergometer used by Dr. Krogh in studying muscular work on carbohydrate and fat diets.

Fig. 98 – Spirometers and a simple kymograph devised by dr. Krogh.

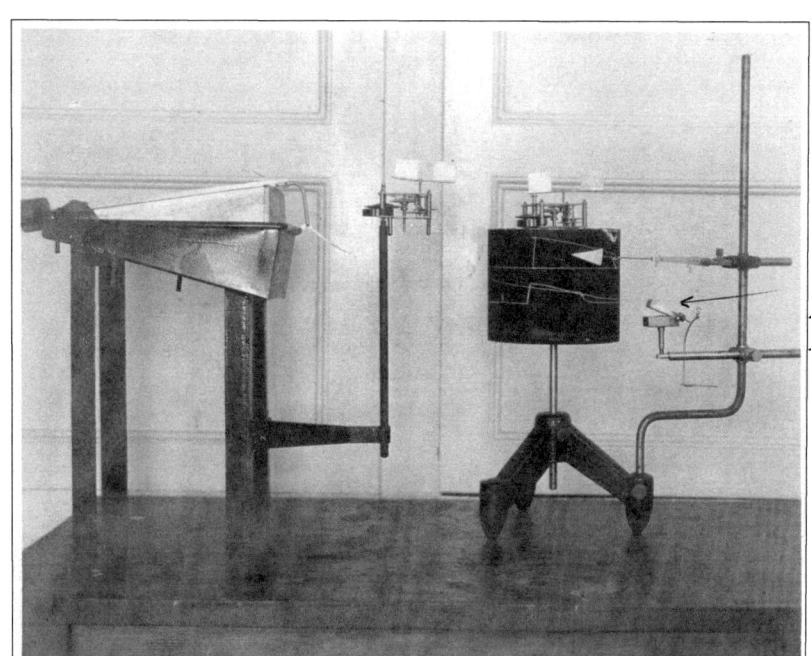

See the very small spirometer for study of insects.

Professor August Krogh, photo taken 1918

up to the present.

I noticed that Krogh prepared his own figures and illustrations for publication except the drawings of apparatus. He has had difficulty with the engraver who insists that the curves must be drawn on plain paper rather than on ruled paper, he complains that the rulings always show, no matter if the lines are blue, hence Krogh draws his curves on ruled paper and then taking a plain sheet of paper, traces over and makes his illustrations to send to the engraver. It seemed unnecessary to me.

On the point raised by L. Hill that the metabolism was higher when measured out-of-doors, Krogh does not think it can be. He says, "What would cause it?" He does not see why the French should at present desire to build a calorimeter. He thinks only in cases where the respiratory quotient is about 100 and a very few other points should remain to be checked up by the calorimeter as against indirect calorimetry.

I saw something of the work which Krogh had been doing on the capillary circulation of blood. The photograph shown in figure 9a was taken in his laboratory where he was studying the capillary circulation. He had been demonstrating certain features of this research to me. He had approached the subject sometime ago by reasoning that during muscular work there must be new paths opened for the blood in order that it should absorb more oxygen, hence he began studying the capillaries and their action to see if this was the probable mechanism by which the greater absorption of oxygen was accomplished. He thinks the capillaries are innervated by two sets of nerves. A light scratch or touch on a capillary will serve as a stimulus for it to open but a strong prick or pressure contract it. The blood supply is itself the stimulus to contraction so that the open paths keep changing and thus in turn all the capillary paths get open and get their blood supply. Dale has discovered a substance which will act as a stimulus to open en the capillaries but no one has yet found anything that will close them. They seem to work independent of the arteries. They can be seen and studied at the base of the fingernail if the skin is prepared with a drop of oil and one uses a microscope magnifying about 50 times. He has been trying to arrange a method for photographing the capillary circulation in the web of a frog. (I was able at the time of the Congress in Paris to introduce him to Dr. Bull who gave him some assistance in reference to the technique for making these photographs). Krogh was much interested in the subject of skin temperature, especially in connection with capillary circulation. He has not laboratory assistant except his mechanician and expects to continue his research on the capillary circulation for about two years. This work he can do by himself and needs no assistance for carrying it on. It, therefore, seems to him a favorable time just now to pursue this line of investigation. He was much interested in Dr. Benedict's work with snakes and snake temperature and referred to his own book in which the fact is mentioned that the female python is 12°C warmer when brooding eggs. He hoped Dr. Benedict would get the opportunity to throw some light upon this phenomena. (See his monograph "The Respiratory Exchange of Animals and Man" p123)

On my second day with Krogh, I saw the automatic gas analysis apparatus which he and his mechanician have been developing during the last year or two. This is for use in measuring the carbon-dioxide from factory furnace flues and in other such places. At the time of my visit Krogh seemed especially interested in the possibility of using this apparatus in mines to analyze mine gases and so automatically keep them below the explosive percentage. He was going to visit England very shortly and one of the reasons was to see Haldane and discuss this problem with him. The apparatus is mounted on the side of the wall and needs only a small pipe with running water. It is the flow of water which operates it, causing it to draw in a sample of 100 c.c. of the flue of gas. It gradually passes this sample through the potash into a small spirometer of Krogh's usual type. (See the very small spirometer B figure 98, at the right hand side of the photograph. This one shown is for collecting the expired air of insects. The one used on automatic gas analysis apparatus is about twice the size of this one in the photograph). As the movable part of the spirometer rises, it picks up a counterbalanced arm when the spirometer has in it 80 c.c. The arm carries a writing point and starting at 80, is moved vertically along a narrow drum carrying a rather long length of paper. The amount by which the writing point falls short of reaching the 100 c.c. level shows the amount in percent of carbon-dioxide in the flue gas. The apparatus is accurate to one-tenth of one percent. After the sample has all been driven through and analyzed the level of the water continues to rise and finally operates a syphon causing the sampling tube to be emptied as well as the spirometer. Then another sample is drawn and gradually passed into the potash and so on. Each time the drum is turned around, a small instrument allows a new line to be drawn by the pen. The apparatus makes an analysis each three minutes and is very regular in its operation.

The pen which is employed is original and I thought of especially practical design. It is shown in the accompanying diagram.

Diagram of Krogh's Pen.

The ink as needed passes from the larger receptacle A into B. The capillary C except when moving upward and even for a part of that time is in the ink and keeps in good condition. The capillary tube is carried by the small arm D. It is held in contact with the drum E by its own weight and tendency to swing down on the axis F. Even at its lowest point, the capillary is in contact with the paper of the drum. I asked Krogh "Is the pen original?" He answered "I think so". It is quite typical of Krogh that he should not have said flatly "Yes," in answer to my question. He impressed me as one of the most careful men concerning the statements which he makes. He does not effervesce or demonstrate much enthusuasm either about his own work or the work of other people. He is the perfect embodiment of the quiet, humble, but keen, indefat-igable scientific worker. In conversation and the demonstration of apparatus, he has continually to be drawn out. He never tells you more or demonstrates further than you ask for, not but what he is quite willing to tell or show what he has, it seems to be the make-up of his disposition.

In reference to the automatic gas analysis apparatus, they have taken out a patent in the United States among other countries and have perfected the apparatus not only for carbon-dioxide but also for carbon monoxide. They figure, I believe, that the complete apparatus will cost about 7000 crowns. It has quite a commercial field. The other arrangement now used for a similar purpose have many points of dissatisfaction. The matter is, of course, of much economic importande in factory management. If a fireman uses his boiler so the flue gases show only 5% of carbon-dioxide, he is losing 30% of the fuel value of his coal. The best results that can be gotten are when the flue gases show 14 or 15% of carbon-dioxide.

Krogh showed me his spirometers of various sizes. Two of these are shown in figure 98. A, is one liter and B about 10 c.c. for use with small insects. He said in general this type of spirometer when in working condition will rise and fall, i.e., follow 60 pulsations of air per second. This means that it is very sensitive. He has a large type of cylindrically shaped spirometer built into one corner of the room which he uses in connection with his muscular work chamber and also for standardizing wet and dry meters. It is arranged with an electrical release and has a graphic recording. The counterbalancing is very simple with small weights.

I was quite interested in a very simple kymograph which Krogh has arranged to be operated with a common cheap clockwork. Note figure 98, C, and D. The shaft of the clock on which the main spring exerts a torsion extends a short distance from one side of the works and fits into a receptacle on the end of the shaft which supports the kymograph. When the clockwork is allowed to run, the whole works revolve on this pivot. Some small projections which can be seen on the lower side of the works in D if figure   engage with the kymograph drum which turns freely on the shaft and causes the drum to be revolved. The speed is changed by the fans shown in the photograph. The fans are tilted at different angles and so their effected resistance is changed without changing the different pairs of fans. A wide range of speed is attained with this simple kymograph which I think would be relatively cheap and an excellent apparatus to use in teaching or for much experimental work. One base, see E in photograph, figure 98, serves for the kymograph support and the support for the registering apparatus, thus everything can be moved together very simply. An extension from the base of the larger spirometer A provides a support for one of the simple kymographs in connection with the spirometer.

Anyone interested in the physiology or respiration is certain to be well repaid for his efforts to visit Krogh. He can secure from him a great deal of information which one feels is very straight forward and given without bias.

Dr. August Krogh died on Tuesday 13 September 1949 after being ill for several weeks with cancer.

University of Copenhagen, Physiological Laboratory.

Professor Henriques.

In showing me the laboratory, Professor Henriquws told me he thought there had been no substantial additions to the equipment since the place had been visited by Dr. Benedict, with the exception of an **electrocardiographic** outfit made by Hill of Lund, Sweden. This outfit stands away from the wall in a room devoted to it. The galvanometer, camera and lamp are on rather tall wooden pedestals on a line about five feet from the floor. The lamp, galvanometer and camera stand on three seperate pedestals and one can walk all around each. The galvanometer is a very good one probably better in many features than that manufactured by the Cambridge Instrument Company of England. The camera is a falling plate arrangement, quite compact and practical. Recently they have been making extensive experiments on the changes in the electrocardiograms with changes in the body position. They had a tilting table to which the subject was strapped, the electrodes being in place, and the body could be rotated to any position as would correspond to standing upright or standing on the head. At the same time under all conditions of position the muscles could be quite relaxed. They were comparing also the heart rate under different bodily positions. I saw quite a number of their tracings which were good and sharp.

Henriques had in splendid condition and operation, several sets of gas analysis apparatus on which he was making some comparison research. It was vacation time and he and his family were occupying a small house in the country. He was kind in supplying me with some addresses which I had been seeking. I found that Professor Hasselbach had discontinued activity in scientific work since he recently had inherited a large fortune and was living in the country, the administration of his own finances requiring his time. I was not able to see him. As it was vacation I also failed to see Lindhard, Sorensen and Hindhede.

University of Copenhagen, Psychological Laboratory.
Professor Lehmann.

Dr. Carpenter, in 1911, had visited Professor Lehmann and Dr. Olsen at the Psychological Laboratory at the suggestion of Professor Krogh to see something of their work on the subject of mental effort and metabolism. I did not feel warranted in asking Krogh to try to make any appointment for me. He was busy working on the proof as I mentioned and was planning a trip to England within three or four days of the time of my visit. It was most generous of him to accommodate his plan to seeing me on the two days that I was at his laboratory and I was very glad that I had written ahead and made this arrangement. I went twice to the Psychological Laboratory, the last time finding an assistant who was also doing the janitor work in this laboratory. He told me that Professor Lehmann had just started on his vacation but he himself showed me through their laboratory rooms although he could not explain the apparatus. They have a large number of rooms, well equipped, and a very good departmental library. I was especially interested in looking over the group of books which had been written by Dr. Lehmann, knowing that Dodge thinks so highly of his work. I saw the room in which their metabolism researches were carried on and found the things exactly as described by Dr. Carpenter in his report of 1911, see his page 40. They were apparently continuing the same line of work. I only regret that I had no opportunity to meet Lehmann and discuss their researches with him. Lehmann's publications are all in Germann. He does not speak English.

Veterinary High School of Copenhagen.
Drs. Mollgaard and Anderson.

My object in visiting the Veterinary High School in Copenhagen was to see a new and very splendid equipment for working in respiration physiology with farm animals particularly milch cows. The apparatus has been fully described under the title Petten kofer Respiration Apparatus in a monograph written by Mollgaard and Anderson in 1917. They gave me a copy of this publications together with two other papers written by Anderson. In connection with describing the apparatus used in the Durham, N.H. work by Drs. Benedict and Ritzman, they had already secured the larger monograph but I was not aware of this when in Copenhagen. The apparatus is very fully illustrated in their monograph by use of diagrams and half-tones and I need not make an effort at extensive description here. The whole equipment is very splendid indeed and gives one the impression of having been built for long continued use. The building was made especially for the purpose and at the time of my visit they were building an enlargement for an extension of this type of work. The whole place was immaculately clean and gave me the impression of a superlative dairy equipment of some sort. The machinery is large and of very rugged construction but is exceedingly quiet in its operation. The ventilating pumps shown in figure 10 of the Mollgaard and Anderson monograph are mounted in a heavy iron frame approximately six feet high and twelve feet long. Something of the relative size can be gauged by comparing this figure 10 with the figure 99 in this report, which shows Drs. Mollgaard and Anderson photograph standing in front of these ventilating pumps. They are both men of tall stature, about 6 feet 2 inches. Their arrangement for feeding the cow during the experiment, for milking it also for disposing of the excrements were exceedingly ingenious.

Up to the present they have used one experimental animal which has been especially provided with a fistula through which they can take off the carbon-dioxide produced in the intestines, due to the ingestion of the grass and dry food. This cow which seems in the best of health and condition has been used by them for more than five years following this operation of placing the fistula. I understood that they had made many experiments, varying the diet of the cow and now could work with other animals and from the constants which they had obtained would know how much of the carbon-dioxide was produced in the intestines of the cow under a certain diet, and therefore the other subjects used in the chamber would not have to be provided with the fistula. I was told their experiments were usually forty-eight hours long and that the animal is in the chamber continuously for this length of time. This gives them a better picture of the actual effect of a certain diet.

The photograph shown in figure 100 was taken in an adjoining room which is a combination of office and recording room for the large respiration chamber. Mounted on a panel board they have a very elaborate electrical recording outfit with fountain pens marking on moving paper registering certain events. There were several wet meters connected with the apparatus and many U-tubes for collecting carbon-dioxide for weighing. Also gas analysis apparatus was employed to supplement the determination by weight.

Mollgaard was at that time extremely busy with government work of an administrative character which had to do with rehabilitating the agriculture in the section of Danish territory which came to that government as a result of the recent Treaty of Versailles. It was just the merest chance that I was able to see him at his laboratory. The apparatus had previously been demonstrated to me by Anderson with whom I had a longer conference. Evidently they have all the money that they can use for this type of work and I was extremely well impressed with their equipment. The man who had charge of their experimental animal and the upkeep of the apparatus seemed to me efficient and careful and I should imagine that there will be an important output from their laboratory.

Fig. 99 – Drs. H. Mollgaard and A. C. Anderson photographed by the large ventilating pump of their respiration apparatus.

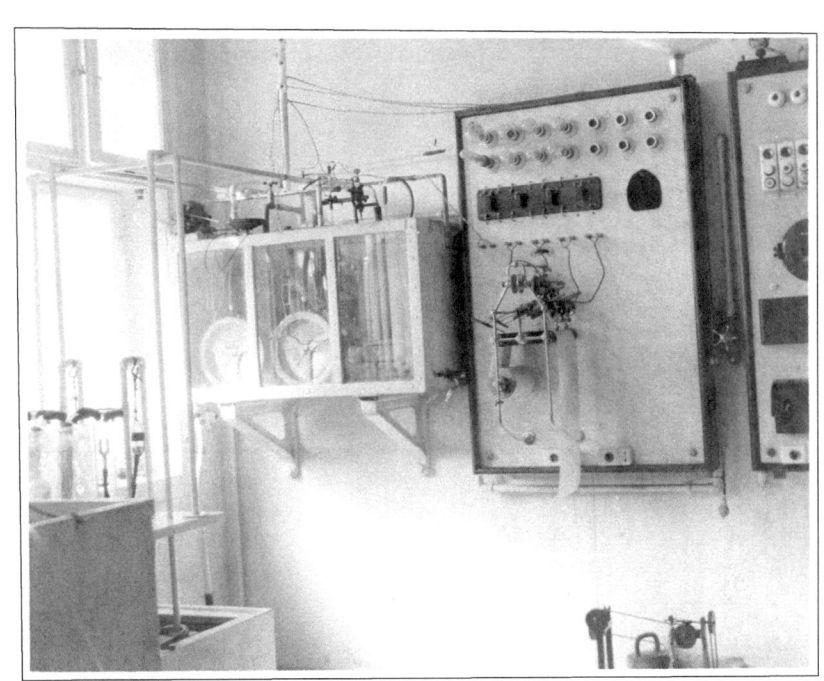

Fig. 100 – Recording apparatus connected with the Mollgaard-Anderson respiration chamber for cows, Copenhagen.

Fig. 101 - Dr. Erik M. P. Widmark in his laboratory working with his distillation apparatus for determining the amount of alcohol in urine, blood or other fluid.

Dansk Maalerfabrik of Copenhagen.
Dr. Gjellerup.

The Nutrition Laboratory has extensively used and recommended to others for use in respiration work the Bohr wet meters which have been manufactured by this company. Recently one of these meters was secured by a collaborator, Dr. Talbot, at a price of $125, approximately four times its price before the war. I asked Dr. Gjellerup concerning this matter and he informed me that they could not afford to sell meters in the United States at a less figure. Their raw material they had to procure from the United States at a price four times the original value and the labor charge is more than four times in advance of what it used to be. He said that if we could not understand these changes or pay the present price he naturally could not be interested in our business. I did not learn what the price would be to an individual or school in Denmark. Their factory seems to be running to capacity. He showed me bills to other customers in the United States and these were at the same rate as that made to Dr. Talbot.

Karolinska University of Lund, Bio-chemical Laboratory.
Dr. Widmark.

After the death of Professor Bang (1918?), Dr. Widmark, who had previously been an instructor in the department of physiology, i. e. in the laboratory under the direction of Professor Thunberg, was put in charge of the Biochemical laboratory temporarily but not elected professor of the subject. In Sweden they have something of the same arrangement as in France; anyone may become on his own initiative a candidate for a certain vacant position. Dr. Widmark has attended to his candidacy as was befitting under the circumstances and, as I recall, there was no other Swede who entered the race against him. However, there is a German Professor, L. Michaelis, trying for the place. Under the present post war conditions many German professors find it extremely difficult to make a living in the position which they formerly occupied. This German is an older man than Widmark and has prehaps more work to his credit, although Widmark has turned out an immense amount of work for a man of his age. It seems that curiously enough the Medical Faculty at the University of Lund prefer the German professor, thinking that he will add considerable strength to their faculty. On the other hand, the Swedish newspapers strongly prefer not importing a German professor to take this position in a Swedish University. They say that the number of Swedish Universities is not large hence the number of positions in comparison to the number of young Swedish scientists is not more than adequate, even if all the positions were filled with Swedes. They publish in large headlines that if one German professor is now allowed to come in and take this position at the University of Lund and displace a Swedish doctor, it will be the beginning of a German invasion of such professors, many of whom are now seeking positions elsewhere and particularly attracted to positions in the neutral countries. The filling of the Professorship has been occupying a great deal of attention of the Swedish public and just how the matter will turn out seems to be somewhat in doubt. I spoke with Professor Arrhenius about it expressing my appreciation of the work of Widmark and hoping he would be elected professor at Lund. I was glad to learn from Arrhenius that Widmark would likely finally be selected to fill the place because of the strong sentiment on the part of the Swedish public.

The building now occupied by this department and the equipment of it is very old and inadequate. When the matter of a Professor has been settled, a new laboratory is to be built and equipped.

In reference to the scientific work of Widmark, my interest was especially in his contributions to the subject of the determining of alcohol in urine and blood. (See fig. 101) He has demonstrated that after a dose of alcohol, the concentration is the same in the blood and the urine. He has modified the Nicloux method for determining alcohol in small amounts in the urine and has developed a very useful technique which can be applied in alcohol experimentation. Widmark thinks that in all alcohol experiments of a physiological or psychological nature, samples of the urine should be taken and analyzed for concentration of alcohol. He said that according to his method, he was able to make twenty determinations, i.e., an analysis of about twenty samples in three hours. He spoke of there being rather wide individual variations in the amount of alcohol found in the urine after a stated dose and thought that this difference is found to be consistent between subjects and should if possible be related to other physiological or psychological changes, to see if there are individual differences in these effects in the same direction. He made the suggestion to me that one should determine the amount of alcohol which it is necessary for him to take every fifteen minutes in order to keep his concentration at a certain point. Then under such conditions of dosing, the physiological and psychological effects might be determined; thus the peak of the effect as usually obtained when a single dose is given would not be such a short time and more measurements could be taken when the same concentration of alcohol was in force. Widmark had shown that the concentration of alcohol in the urine is independent of the dilution of the dose and that also after a dose of alcohol has been taken, if later, by a half hour or an hour, a liter or more of water is drunk so increasing largely the quantity of urine excreted the concentration of alcohol in these latest samples of urine is such as to follow the curve that could have been expected from the original dose of liquid containing the alcohol. In other words the concentration is not decreased by adding water and so increasing the volume of the urine. Widmark, so far as I could learn, has made no attempts to correlate the measurements of alcohol effects with concentration in the urine. He thought this line of research probably a very profitable one. I am very glad that due to the kind cooperation of Dr. Carpenter, it was possible for me to combine this line of observation with the alcohol experiments on 2.75% alcohol solutions performed in the winter of 1919-1920.

Sweden

Karolinska University of Lund, Physiological Institute.
Professor Thunberg and Dr. Westerlund.

In accompany with Dr. Widmark, previously an instructor in the department of Physiology, I visited that very well arranged and equipped Institute. Professor Thunberg was away it being the time of vacation and I had no opportunity of meeting him while I was in Europe. (I am glad to record the pleasure of his visit at Nutrition Laboratory Oct. 21, 1920.) I believe the Institute was planned and built by Professor Blix, a very original and ingenious man. The original features from the especially designed door-latches by Blix to the large variety of physiological apparatus designed and constructed in the shop of Sandstrom, which shop is now in charge of Hill. They have, and in very good working order, several of the Blix-Sandstrom kymographs, the electrocardiographic apparatus, manufactured in that shop.

A thing which impressed me in the department of physiology was that Thunberg for a number of years has given a short course in statistical methods as a part of the course of physiology. The students are made to use these methods in dealing with the data which they collect in their practical experiments, e.g., blood pressure measurements, blood counts and all other experiments where the results are quantities and the experiments are several times repeated. It seems that Thunberg adopted this plan at the suggestion of the local astronomer, who showed him not only how important it was to consider things in terms of averages but also in relation to their variations from the average. They had a short syllabus for this subject published in Swedish.

I visited the machine shop of Hill which is behind and near the Physiological Institute but, I believe, is not owned by this Institute. Professor Brielaus of the Physics department showed me the shop and their methods of work and product. They are making besides the electrocardiographic outfit, other recording apparatus, a vacuum pump, thermostats, electric centrifuges, etc., etc. The prices seemed to me rather high and I did not see anything that I thought was particularly needed by the Nutrition Laboratory at present. As recorded in connection with my report on Salomonson's laboratory at Amsterdam, (See fig. 76), Hill is making an episcotister using the regular base of their electrical kymograph as the driving arrangement.

Karolinska Institute of Stockholm, Physiological Laboratory.

Professor Johansson.

The circumstances of war which naturally produced many problems for neutral nations have so shaped conditions that Professor Hohansson has been as busy as ever with commissions and boards and hence the amount of experimental work which he has done in recent years is practically nil. He is a ve very charming personality and, of course, well fitted for work which brings him into contact with public officials. A satisfactory photo of Hohansson at his desk is shown in figure 102.

Since the visit of Dr. Benedict in 1913, they had constructed a new respiration chamber in the depression made in the floor in the basement room where was located the respiration chamber which Dr. Benedict speaks of, and illustrates in his figure 90, which was modeled after the bed calorimeter in the Nutrition Laboratory. This new chamber is made of sheet metal and so arranged that the upper part can be raised to allow the subject to get in and out. In figure 103, which shows a general view in the basement laboratory, Johansson is standing at the left, the assistant is at the endpf the room and the mechanician, Mr. Yarl, is in the new respiration chamber with the Johansson Ergometer.

The Johansson Ergometer has also been constructed within recent years and is shown in two photographs, figures 104 and 105 in which Johansson is in position for working the ergometer. It is arranged so that both positive and negative work may be accomplished independently. I thought, however, the working position was not particularly favorable. If one leans against the table with the chest although the table is shape. for the body, still there is apt to be some hampering of the respiration and one cannot easily divorce the work with arm muscles and back muscles. There is an indicating arrangement shown in the extreme right of figure 164 by watching which the subject

Fig. 102 — Professor J. E. Johansson at his desk, Karolinska Institute, Sweden.

Fig. 103 — General view in Johansson's laboratory for muscular work and respiration measurements. Prof. Johansson standing at the left, Mr. Garl, mechanician at the ergometer.

Fig. 104 – Professor Johansson using the ergometer of his own design.

Fig. 105 – The Johansson ergometer is placed inside a respiration chamber special accommodation for which has been made in the floor of the laboratory.

can continue to do the same amount of work at the same rate. The cover of the chamber when lowered goes into a water seal as can be seen at the lower part of figure 105. The approximate size of the chamber may be guessed from these views which show both mechanician Yarl and Professor Johansson inside of it. The chamber is barely large enough for using the bicycle ergometer. The equipment for research here is very satisfactory and it is regretable that it is not being used more nearly up to capacity.

Johansson's Laboratory was in splendid order. He has much ability and interest in having a place for everything. The instrument cases are very numerous and contain grear quantities of smaller types of physiological apparatus. There are many things of an historical value and a careful catalogue and description of these things is in preparation. He had continued his accumulation of lantern slides made from original sources. These he uses very largely in his teaching work, and I imagine, with very good results. His niece, Miss Brita Johansson, is his secretary and photographer.

One of the especially pleasant features of a visit to Stockholm is the opportunity of seeing Professor Arrhenius, Director of the Scientific Academy of the Nobel Institute and worker in Physical Chemistry. Professor Cohen of Utrecht had been kind enough to suggest giving me a card of introduction to Professor Arrhenius. I had no need of this card, however, being in company with Johansson and although Arrhenius was strenuously employed at the time of my visit in looking after the American-Swedish singers who were then in Stockholm he found time to come to the home of Johansson for dinner one evening. (See figure 106). It was delightful to find how interested he was in the Nutrition Laboratory and in American Science in general. His genial conversation quite captivated me. I had a little unexpressed wich that before long he would turn his mathematical skill to some consideration of the effects of alcohol.

Stockholm Board of Health, and Technical High School.
Professor Sondén.

In the Board of Health offices which I visited in company with Professor Johansson, Dr. Sonden had been occupied during the recent years with the routine problems of the city's sanitation. I judged that the larger portion of his time during the last three or four years had been devoted to the new Technical High School of Stockholm. He was a leading member of the committee who had in charge the planning and building of this huge institution. They had practically completed the building in the very early part of the war before there was much change in prices of building materials and labor. It was for them a most fortunate thing that they got this Institution built and equipped at that time. It is a very splendid structure appropriate as a great public institution and practical in the facilities for teaching. Sonden spent a half day in conducting Johansson and myself through the various departments of this school. The equipment for chemistry, physics, electrical engineering, mining engineering, steam engineering, hydraulic engineering etc. seemed to leave nothing to be desired. The drinking fountains, the lavatory facilities and all such things were entirely designed and planned by Sonden, who occupies in the school the position of professor of hygiene. He showed us many interesting, ingenious teaching devices which he had for illustrating problems in reference to ventilation, furnace gases, oxidation of oil and coal and other fuel materials, purification of water supply and other problems that naturally are presented in such courses. The photograph, figure 107, was taken in Sonden's office in the new Technical High School. The picture on the wall is of Berzelius. I think that Sonden now has a much broader field for his noteworthy abilities than he probably had previously in the Stockhlm Board of Health. I understood that he would continue to play a part in that organization which will soon have a new building, there being plans to tear down the present building.

Fig. 106 - A group at the residence of Professor Johansson 17A Kungsklippa St., Stockholm.
Sitting :- Souden, Miles, Arrhenius, Johansson
Standing :- Stenström, Strömbeck, Miss Johansson, Miss Geiger, N. Johansson.

Fig. 107 - A group in Professors Souden's office in the new Technical High School, Stockholm.
Souden, Miles and Johansson.
Portrait of Berzelius

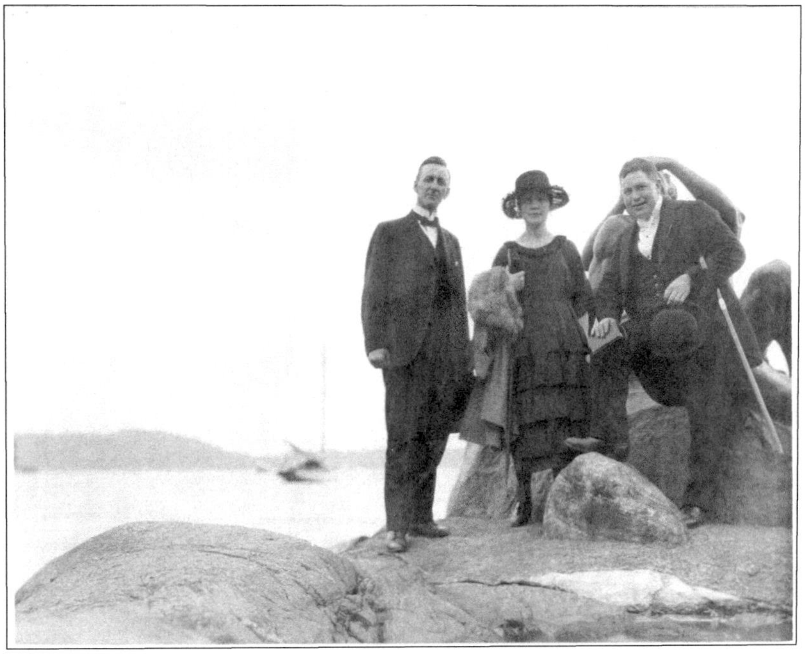

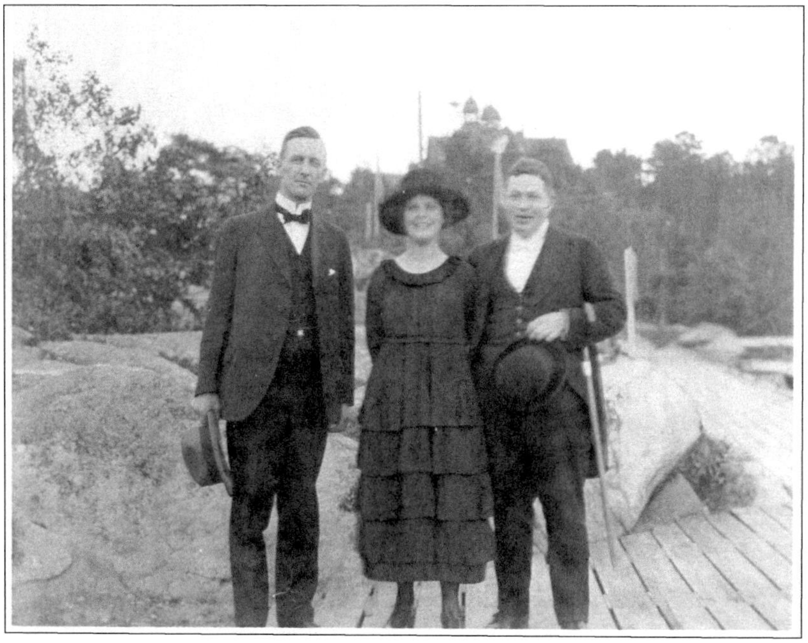

228

Stockholm
Den 4 Juli 1920
Kungsklippan 17

J. E. Johansson
Svante Arrhenius, f. 1859.
Klas Sondén
Nils Stenström
Jan Paul Strömbeck
Nils Johansson
Meyer Geiger
Brita Johansson
Walter Müller

This party card belongs with Fig. 106, p 227.

Nils Stenström
1920.

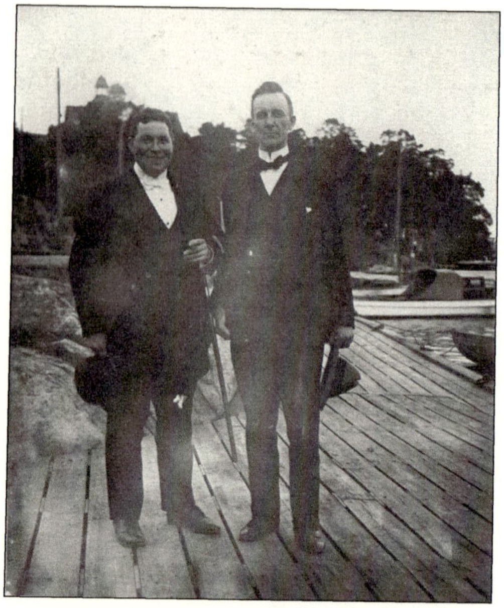

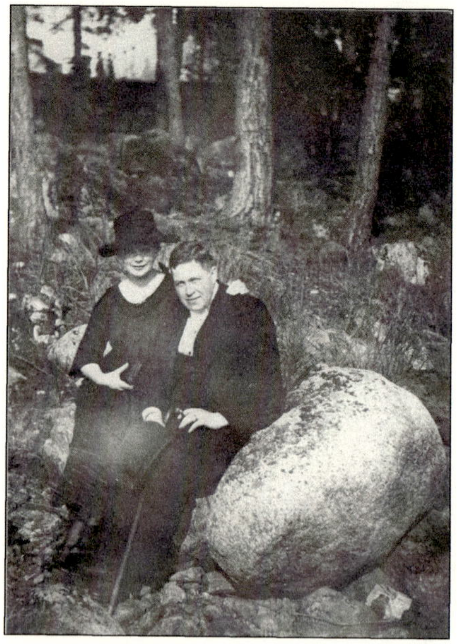

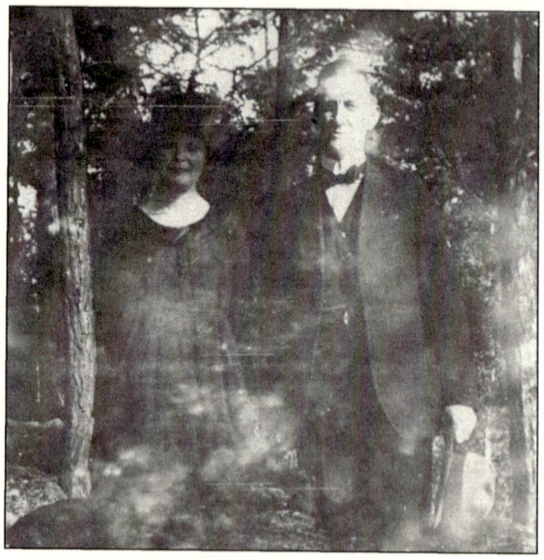

Garnisonssjukhuset Milit.

(Military Hospital of Stockholm.)

Dr. Stenström.

It was very pleasant to meet Dr. Stenström (See fig. 106) in Stockholm as he had visited the Nutrition Laboratory for about two weeks in March of 1920 at the same time that Professor Van Leersum was here. His research interest is respiration physiology as illustrated in some papers which have been jointly produced by himself and Dr. Liljestrand. Recently, however, he has given special consideration to the study of the electrocardiogram and before visiting America had spent two months in the laboratory of Einthoven and about a similar period in the clinic of Lewis. He has at his disposal in Stockholm at the "Serafimerlasarettet" an electrocardiographic outfit made by Hill of Lund and he proposes beginning research work with this immediately. He had just recently arrived from America having been over here much longer than he originally expected due to repairs necessary on the ship. He came from Stockholm to America as the ship's doctor on The Stockholm of the Swedish-American Line and remained in this country six weeks while the ship returned to Stockholm and came back. This arrangement of serving as ship's doctor and so having an opportunity of visiting America for scientific purposes is a very good one. I was able to recommend it in Holland and am sorry to learn it may be discontinued in Sweden.

Stenström is a bright fellow, and has done remarkably considering his age. He will continue to do research with Liljestrand and hopes soon to change from the military Hospital which position he now occupies for the purpose of making a living, to some other place in Stockholm where he can devote more time to research. He is a wonderful swimmer and can endure long exposure to cold temperature. While at the Nutrition Laboratory he kindly coöperated in lending himself as a subject to Dr. Benedict and myself for some measurements on temperature after prolonged exposure to cold air. He has served as an assistant in Johansson's laboratory and thought it was regretable that at the present time so little work was in progress there. He said that Johansson was glad to have a sample of all kinds of apparatus for demonstration purposes but had no time for carrying out researches himself. Dr. Stenström was most kind in showing me about Stockholm and some of its environs.

Germany and Austria

University of Hamburg, Physiological Laboratory.
Professor Kestner.

It was convenient for me to pass through Hamburg, Germany, en route from Holland to Denmark. I found Professor Kestner, (formerly Professor Otto Conheim), at a small temporary laboratory in connection with the Ependorf Krankenhaus. Just prior to the war the University of Hamburg was established and Kestner was made professor of physiology and he planned and began to build a new Institute for Physiology. The building was hardly complete at the beginning of war, i.e., tiled floor floors, plumbing and inside fixtures were not complete and the Institute was wholly without scientific equipment. During the war it was used for temporary hospital purposes and in June the German physiologists held their meeting here. Kestner seemed to me very much depressed. There appeared no probability that the Institute could be equipped for a number of years. It would literally cost millions and millions of marks at present to equip it for teaching and research. He having visited Nutrition Laboratory, was interested in hearing about its recent work and also about Dr. Benedict. It seems that in January or February, 1919, Kestner published in the Hamburger Fremsterblatt and also in the Frankfurter Zietung, "An Open Letter Addressed to Dr. Francic G. Benedict, Director of the Nutrition Laboratory, Boston." The letter concerned the under-nutrition of the German people and was something of an appeal. He was unable to supply me with a copy of this letter, but I believe the matter has been brought to the attention of Dr. Benedict.

I found some small amount of experimental work in progress in his small laboratory which is a one story building and of rather temporary construction. He was doing some respiration work with animals. The apparatus was extremely meager and mostly home made. He had one small Williams bottle. He said that every day he reminded his assistants to be careful in weighing it. The glass intake tube had been partly broken of. He did not know what would happen to his research if this Williams bottle was to get broken. I told of the large supply which I had

234

recently seen in the Laboratory of Professor Buytendyk and suggested the possibility that he might obtain one or two from him. Kestner had two dogs prepared after the manner of Pavlov for the production of gastric juice and for experimentation of this character. The dogs seemes to be in fairly good condition. Kestner was one who said that at the Scientific meeting, they could of course meet with the Americans they could, they thought, meet probably with the English, but with the French, they could never meet. He sited the situation at Frankfurt and the French use of the colored troops. We did not talk about the war. I was not entirely certain that he was very glad to have me come. He told me of their need of scientific literature, of the kindness of Van Leeuwen and showed me his new building.

University of Berlin, Physiological Laboratory.
Professor Rubner and Dr. Gildmeester.

On returning from Sweden towards the Physiological Congress in Paris, I passed through Germany and stopped at Berlin for five days. (I may record a part in my conversation with Professor Heger when at Brussels. He asked me if I planned to attend the Congress in Paris. I assured him that was my intention and that between visiting him and the Paris Congress, I was going to Denmark and Sweden. He then said, "Well, that means that you will visit me again on your return trip for Paris. It is impossible that you can go to Paris any other way than to return through Belgium." It seemed to him quite unthinkable that I should pass through Germany. I knew from my Quaker acquaintances that it would probably be quite possible for me to pass through Germany without any trouble, even though the United States had not yet made peace with Germany. From the German Ambassador in Amsterdam, I had no difficulty in securing a vise of my passport for Germany particularly as I am a Quaker and the work of this group in Germany at that time was deemed by them of great help and importance.) While in Berlin, I took the opportunity of making a call on Professor Rubner at his home. It was examination period and he said there was no time when he could meet me at the physiological laboratory. He probably is a man who always seems a little distant and formal on first acquaintance. He told me somewhat about their condition, how impossible it was to do any scientific work because always there was the problem of how to secure more food. It was a pressing and disturbing problem in his own family and absolutely disorganized any abstract scientific work. He expressed interest in the under nutrition research carried out at our Laboratory and asked to see photographs of the Laboratory and the workers. He told me something of the building of the Institute for Arbeits physiologie of which he is also director, Dr. Karl Thomas is directly in charge of this Institute. Rubner expressed no animosity towards the Americans and but very little towards the English, but quite the reverse in reference to the French.

On the following day, I visited the Physiological Laboratory about which I was courteously shown by Dr. Gildmester. This Laboratory and its equipment is familiar to my colleagues at Nutrition Laboratory. I learned that the equipment had not been supplemented during the war and most of the apparatus looked old and not in very splendid condition. There was, however, a tremendous amount of it and a great number of rooms, all of which seemed thoroughly well crowded. It was examination period, as stated, and I saw no students working and no research in progress of special interest to us. Gildemeister has the work in electrophysiology. He said that he could not consider the resumption of the Zeitschrift f. bilologische Technik und Methodik due to the heavy cost of printing, to the fact that there was not a great deal of research going on and that the Germans were at present more or less cut off from the scientific world. I saw here a most marvelous collection of large teaching charts for physiology, many of them colored. A group of rooms in the attic were entirely devoted to storing these charts in a classified order, together with one room for drawing and photographic work in preparation of other charts to keep the collection up to date.

University of Berlin, Institute for Arbeitsphysiologie.

Professor Rubner and Dr. Thomas.

Dr. Gildemeister took me to this new Institute which is only a few steps from the Physiological Laboratory and introduced me to Dr. Karl Thomas who is in charge of it. The Institute was built in 1916 and although the war was in progress, its construction is quite satisfactory. It is not as large as the Nutrition Laboratory building but about that size and shape of a building. One little point of construction noticed was the upper panel of all the inside doors were blackboards with a little chalk rail below. The rooms are not large. There is provision for a number of workers but although built for the study of problems of the physiology of work, the equipment is almost entirely of chemical nature and to date they have been interested in problems of food and the chemical analysis of food. They were working on the matter of the utilization of old potatoes when I was there. Thomas seemed appreciative of my visit, although he showed me about rather hastily, all the time apologizing that he had not much to show. He is a man of very quick movement and speech but very pleasant. He spoke much of their need of literature, that they found it impossible with the present rate of German money, to buy foreign scientific literature or even to subscribe to the scientific journals. He desired to have me arrange an exchange for him, supplying him with a subscription to the Journal of Bio-chemistry and he would supply me with a subscription to a new fortnightly issued abstract journal of physiological literature. He gave me a very generous supply of reprints. I met no one else who has been connected with this laboratory. Dr. E. Weber has worked there and I should have been glad to have met him. I tried but was not successful, due to the holidays, to visit the Laboratory formerly directed by Dr. Zuntz at the Agriculture High Shool and also the Physiological Laboratory of Dr. Max Cremer.

While in both Hamburg and Berlin, I inspected some of the work done by the American Friends Service Committee in feeding of German children in connection

239

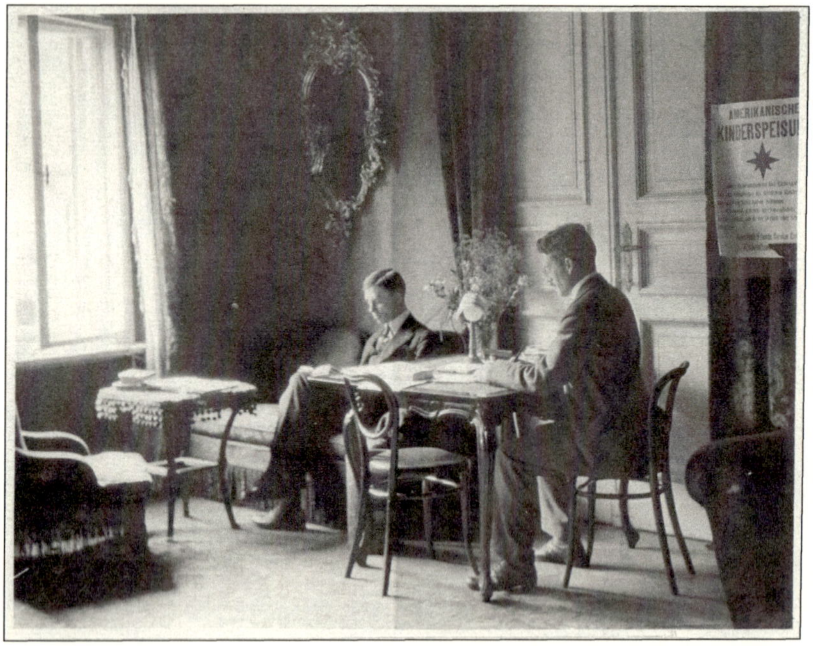

with schools. I was very well impressed with the quality of the food supplied and with the method of distribution also with the way in which the children used the food. Nothing was wasted. There was a clearest possible indication that the food was needed and was greatly appreciated. In July they were feeding about 700,000 per day, giving them the midday meal.

University of Leipzic, Institute of Experimental Psychology.
Professor Kruger and Dr. Kirschmann.

Dr. Kirschmann, a German by Birth, was for several years professor of phychology at the University of Toronto, Canada. About 1913 or 1914 he had an illness in the nature of a paralytic stroke and was given leave of absence for recuperation. He went to Germany and happened to be there at the time the war began. It was naturally impossible for him to return to Canada as it was also undesirable on his part. With the continuation of the war, the University of Toronto dismissed him from the faculty. His support was thus cut off and Professor Wundt, then director of the Physiological Laboratory and greatly in need of an assistant due to the fact that the younger men had been taken by the government, asked Kirschmann to come to the Laboratory as an "ober" assistant, the same position which he had occupied as a young man in 1893.

Professor Wundt, who is now 88 years old, has retired and is living in the country. The new director of the Laboratory is Professor Kruger, whom I met and with whom I had a very brief conversation. Kirschmann seemed pathetically glad to have a visit from an American and he most obligingly I might say, insistently, showed me over the whole Laboratory of Psychology. Naturally I was interested to visit the place where so many Psychologists had at one time or another been students. Although the building is relatively new, that is, it is not the building which housed the original Psychological Laboratory, I was rather astonished to find that everything was most severe. There was nothing in the nature of comfort indicated in the chairs, which furnished offices or in the desks or tables. The rooms seemed to me rather narrow and crowded. The office of Professor Kruger which was formerly the office of Professor Wundt contained only a few books and the barest necessities. Kirschmann said it was the same way when occupied by Professor Wundt. Kirschmann showed me how he was able to imitate any metallic luster by combining different thin layers of slightly colored gelatin. He was especially interested

GERMANY AND AUSTRIA

242

Image not reproduced

Image not reproduced

Image not reproduced

in psychological problems of vision and light. A research student under his direction was working on the effect of different colored lights when the light was so arranged as to practically occupy the whole field of vision of the subject. The subject sat behind a large white sheet, a projection lantern on the other side was used as a source of light, the light was passed through filters of different colors and hence the subject who sat near the sheet behind had practically the whole field of his vision covered with this particular color of light. Under these circumstances, they put him through a number of measurements. Kirschmann showed me a stereoscope of his own design for experimental work by which he separated the two pictures mounted on separate cards the two pictures which were to be combined in the stereoscopic image could be tilted at different angles and given different degrees of rotation to see how far these various factors could be extended without destroying the stereoscopic appearance. It was a nice instrument and showed the stereoscopic image to be remarkably tenacious. As I recall he said he was the first one to record the nature of the eyemovements during reading but had never received credit.

While visiting the laboratory under the direction of Kirschmann I was introduced by him to Professor Wirth, director of the Seminar for Psycho-physics. He spoke of just having received what he thought was a very nice letter from Professor Spearman of London. I could see clearly how much he appreciated getting into touch with Spearman again.

University of Leipzic, Physiological Institute.
Professor Garten.

For me it was of very great interest to visit this Laboratory which has been made famous through the directorship of Ludwig and Herting. Professor Garten, who has lately come into this position as director of this Physiological Institute of Leipzic was very cordial. He thanked me several times for coming to see him, saying that they felt very much cut off from the scientific world. Nothing was said in our conversation about the war or the Physiological meeting in Paris or that which occurred in Hamburg, but he told me seemingly with the greatest freedom of the researches on which he had been engaged during the war. These had concerned problems in reference to aviation and the selection of aviators. He had developed an apparatus for more or less imitating the conditions of controlling the aeroplane. The subject was seated in a chair which might be revolved forward or backward or sidewise or in any axis whatever about a pivotal point which was a little below the seat. He was tilted and then by a stick control he tried to rectify his position and bring himself back exactly to the first position. The error of his compensation was read in degrees by the experimenter. He put me in the apparatus and made several tests with me. The apparatus is figured and described in a reprint which he supplied. See figures two, three and four in this reprint. I was pleasantly surprised at his seeming willingness to show me anything he had. In the lecture room, they had put up a long swing coming down at the end of the lecture desk and by this arrangement they could quickly test the nystagmus time. The two ropes of this wing were twisted around a certain number of times and in this way a given number of turns could be supplied to the subject and in about a given time. They could also combine the turning with other movements if they desired.

Garten, as is well known, is particularly interested in string galvanometer technique and electro-physiology. He had a very well equipped room devoted to this subject. The room looked like a place in which a very large amount of work had been done. Its equipment seemed exceedingly practical. He had a film camera something like the Frank kymograph. Of this he had several of the same model and used it in connection with all his string galvanometer equipment. His string galvanometers were all of the Edelmann type. For vibrating time marker and to secure a wide aptitude of vibration he used a tuned reed mounted in a pipe and the air was drawn through this pipe by a small pump attached to the faucet. The reed was therefore caused to vibrate before the slit of the camera much like an organ reed. He said he thought the device was original with himself but later found that someone else had made just such a thing. It seems that this former author thought it was original in his case and said so but Garten had found that still an earlier man had made the same device so all three had made it independently which seemed a little odd but probably is the case with a good many scientific appliances and methods.

Professor Herring, the previous director of this laboratory, was particularly interested in the problems of light and vision. There was much demonstration apparatus for this field of work and a particularly large and complicated spectroscope for providing light of different wave length for experimental purposes. In a dark room Professor Herring had arranged a small window so that it might be covered with glasses of different color and the sunlight passing through these screens falling on different backgrounds provided quite wonderful contrast effect, the most beautiful teaching arrangement of this character that I ever saw.

In Garten's office I saw a copy of the United States Government Publication entitled "Air Service Medical". Garten pointed this out to me saying that he knew something of the work in reference to aviation that has been done in the United States. It seems that this book came to him through someone of the neutral countries. He was pleased that I desired to take a photograph of him at his desk in his office. I regret that I had only a vest pocket camera with me at this time. (See fig. 108.)

For photo of Professor Siegfried Garten see page

University of Vienna, University Hospital Kinder klinik.
Professor Von Pirquet and Dr. Schick.

In visiting Vienna, I had hoped to see Professor Exner and also Professor Durig. It so happened that the former who has retired is living in the country quite a distance from the city and the latter had, on the day of my arrival, sent his family away on their vacation and was himself following on the next day. My friend, Mr. Roger Clark, at the head of the Quaker Mission in Vienna saw Professor Durig at the railroad station and told him of my presence and my desire to see him. He begged to be excused on account of his poor health and the fact that he was just leaving the city and said that he was leaving a small book for me which he had lately written.

I visited Dr. Van Pirquet at his home and also at the Hospital at Kinderklinic. On the second day of my visit to the Hospital which visit began quite early in the morning, I had also the company of Dr. Alfred F. Hess of New York. Von Pirquet has recently devised and put forward a new termanology in reference to nutrition. He has two methods for testing under nourishment in people. Naturally the subject of under nourishment is of extreme importance in Vienna at present and has been during the past few years. One could see clearly in passing down a street that the thing the people needed and the thing they were interested in was food. Under the present conditions it is impossible to feed everybody in Austria. The question arises how to sort out the people and select those most under nourished and most in need of supplementary food by relief organizations working in that city. The first method of Von Pirquet is called "Pelidisi" meaning the measurement of weight and length of body compared to the average. Von Pirquet claims that experience has taught that the following equation is correct for grown people. The weight of the individual times ten equals the cube of the sitting height in centimeters. An adult with a sitting height of 90 cm. has for instance a normal weight of 90x90x90 or 72900 grams or 72.9 kilograms. A school child does not have as much fat and muscle as the adult so that the Pelidisi is on the average not 100 but 94.5. Children with a Pelidisi

247

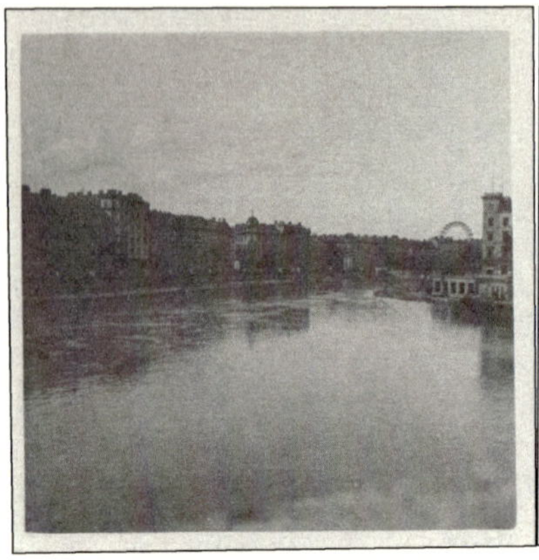
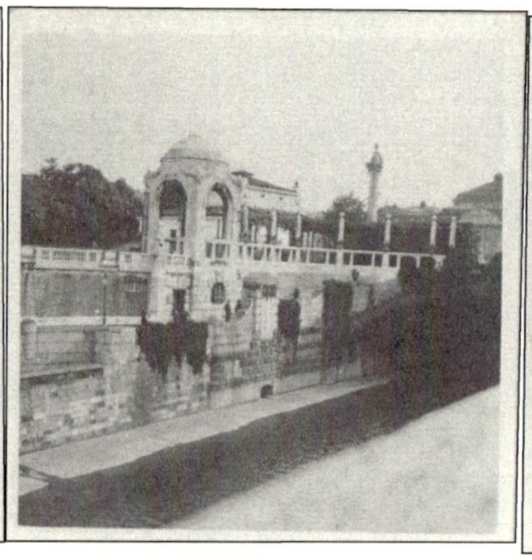
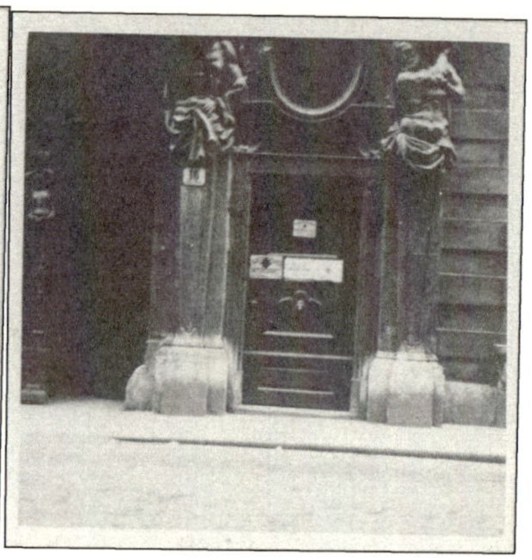
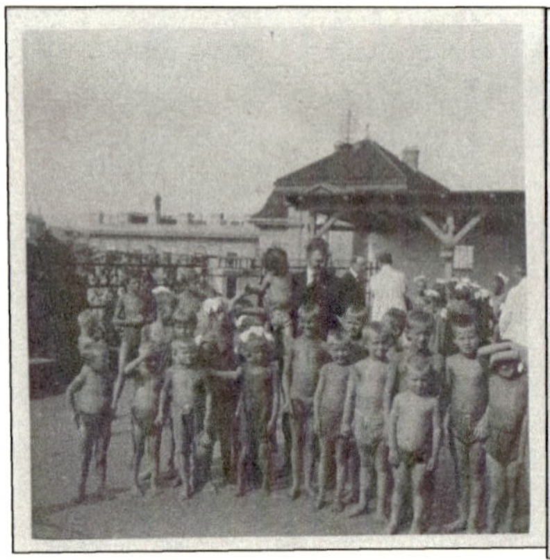
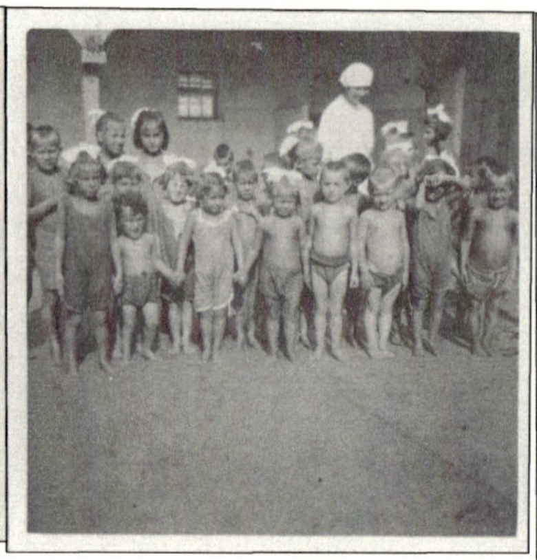
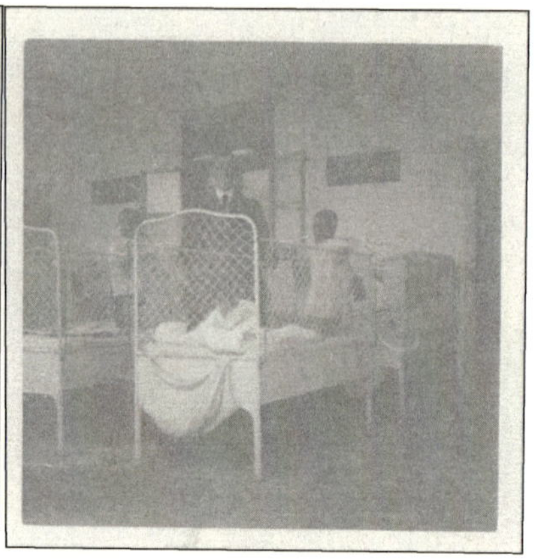
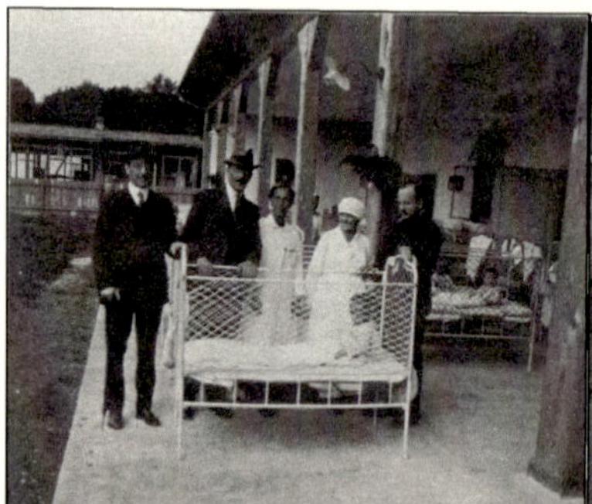
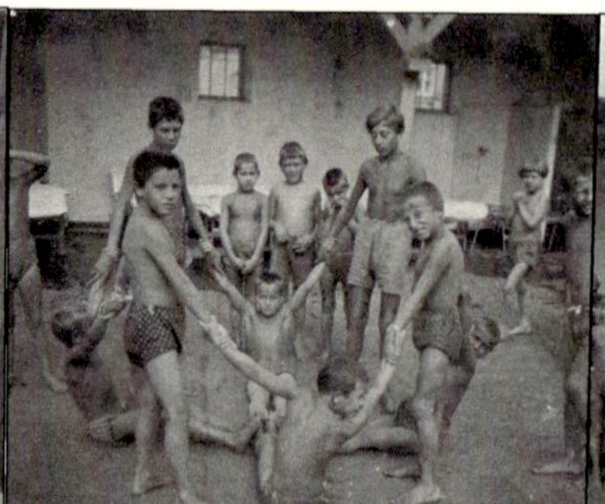
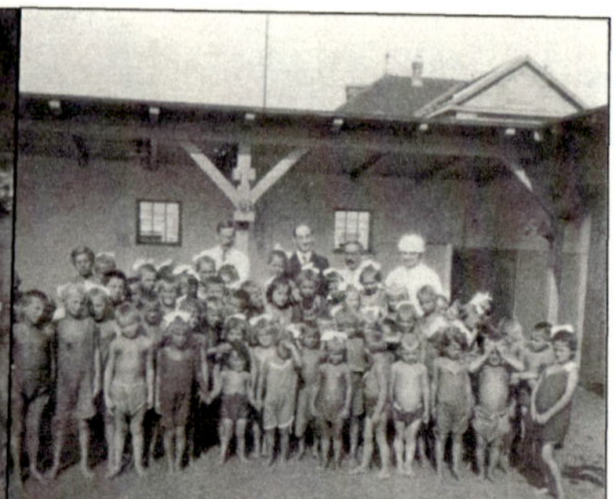

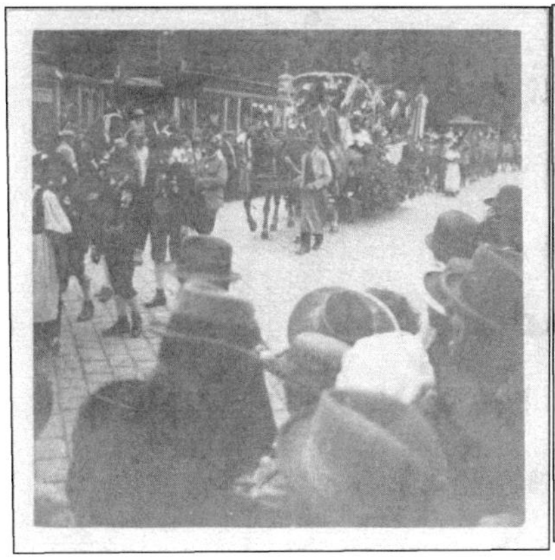

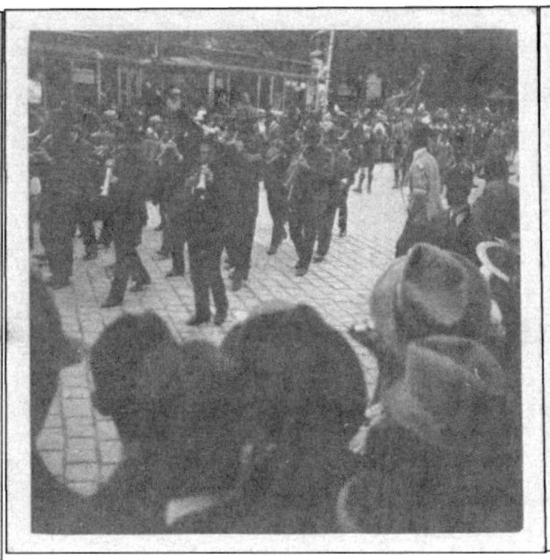

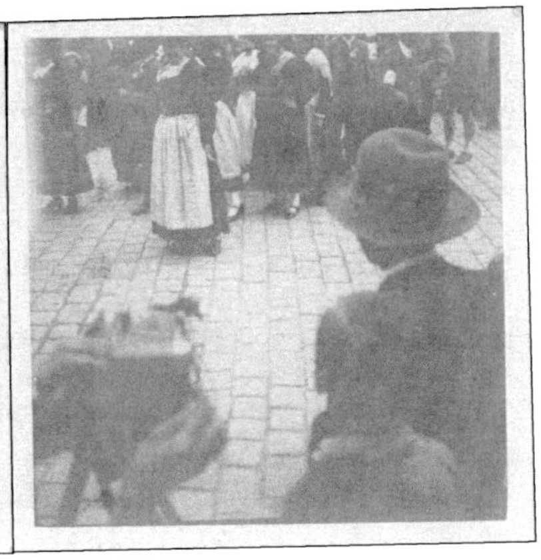

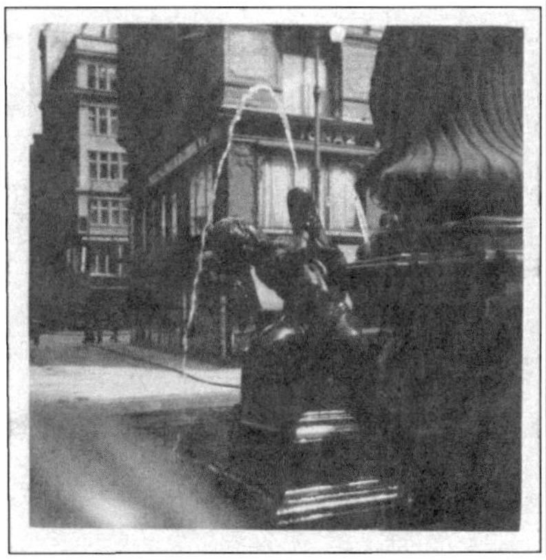

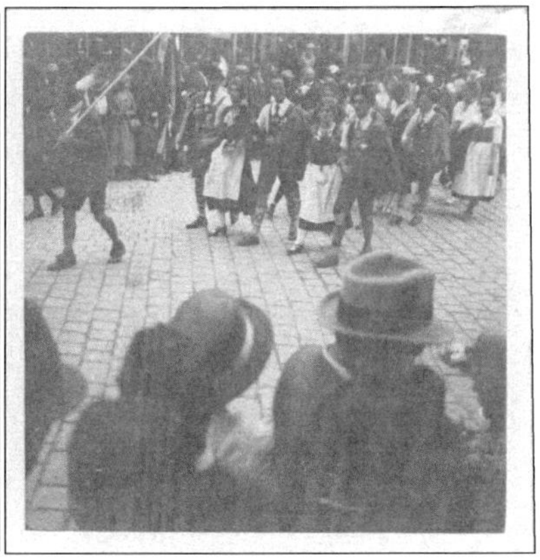

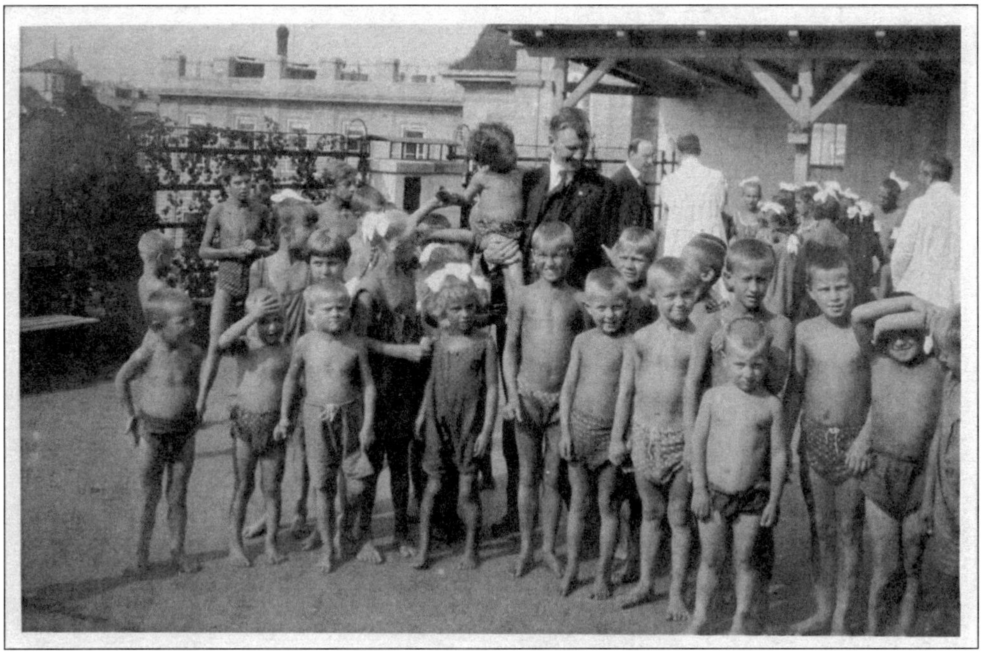

of 94 and less are considered under fed. Those with a Pelidisi of from 95 to 100 are considered well nourished and those with 101 and above as over fed. Inspectors of the American Child Welfare Mission record the sitting height and weight of all children attending school in all towns having feeding stations. The Pelidisi is then found by referring to a table and thus children selected who are to receive food. The selecting is made on this basis of the state of nutrition regardless of the class or income of the patient and without the influence of religion or party political prejudices. The second method of measurement of undernutrition is what Von Pirquet calls "Sacratama" meaning measurement of condition of blood fatness, skin fullness and muscles. The work Sacratama is a composite word in which the consonants or constants indicating the different bodily elements or parts examined and the vowels varying indicating the condition of the different parts. Thus "Sacratama" gives in one word according to Von Pirquet's method, the whole opinion on the state of nutrition. In accordance to this summary, the children are classified into four categories: to class 0 all those children belong whose state of nutrition does not contain a bad mark, that is, no O or U for instance; class I those children belong who have one bad mark, For instance, Sacratama; SO for instance stands for reduced blood coloring of the skin, a pale skin surface; CRE for an abundance in fat; CA for a medium turgor; MU a very weak muscular system. The combination SE-CRE-TA-CUM gives in one word the whole thing and that is the case of nutrition. The word Sacretama would mean a meager but otherwise normal child. Class 2 includes those children who have two bad marks, for instance, Socrotama a pale and meager child; class 3 finally includes those children with three or more bad marks, for instance, Sacretomu, a child of normal complextion but which is emaciated, poor in muscle and has a flabby skin. On the whole that Sacratama examination agrees with the Pelidisi examination. Thus the examination of all primary schools of a Vienna district give the following result.

Pelidisi	100	99	98	97	96	95	94	93	92	91	90	89	88	87	86	85
Per cent of children	.36	.56	1.24	2.2	3.6	5.7	9.3	14.0	14.2	15.0	13.1	9.2	6.0	2.7	1.2	.64

Sacratama classes	0	1	2	3
Per cent of children in each	7.3	4.3	13.8	74.6

The method which they now use in selecting is first to use the Pelidisi in determining all the children of the school. Those children who show 94 or less are forthwith admitted to the feeding. The remaining children are examined as to their Sacratama and those most qualified admitted depending on the proportions of food still at their disposal. Examinations are repeated every two months and children who have regained weight above 94 Pelidisi or have improved their Sacratama are eliminated and replaced by children with bad marks in nutrition.

The feeding of Austrian children with American food is carried out under what Von Pirquet calls the "Nem System". This system takes as its unit of measurement the nourishment value of one gram of mother's milk. This basis value is called "Nem" for the initial letters of the three words Nutrition, Element, Milk. The nourishment value of 100 grams of milk is called hectonem and abbreviated Hn., of one thousand grams kilonem and abbreviated Kn. and of one thousand, tonnenem and abbreviated Tn. All food stuffs are compared in their calorific value with the milk standard. The milk standard equals 667 small assimilable calories according to Koenig, or a large assimilable calorie equals one and one-half Nem. The nourishing value of the American food stuffs according to the Nem system measure up as follows: Evaporated milk, 2 Nems for a gram; One tin contains 450 grams or 900 Nems; Sweetened Condensed milk, 5 Nems and a gram, One tin contains 400 grams or 2000 Nems or 2Kilonem; Cocoa, 6 Nems and a gram; Sugar, 6; Flour, 5; Rice, 5; Peas, or beans, 4; Stock, that is lard or lard substitutes, 13 1/3 to the gram; Bacon, that is pork, 10; Canned meat, e.g., corned beef 2½ Nems in a gram. As a uniform measure for the meal of a school child, the food quantity of 1000 Nems or 1 kilinem, roughly 1 meter or one quart of milk was taken. This food quantity had a caloric value of 667 large assimilable calories or of roughly 700 large raw calories.

I do not know that it is necessary to say anything more in detail about the Von Pirquet Nem system and the way which it is carried out in the feeding of children in Vienna and also its use to some degree in the feeding of adults. Von Pirquet told me that the system of animal nutrition and its nomenclature was in very much better condition that that of human nutrition and that in animal nutrition they made a unit the gram of cows milk. He had tried to do the same thing for human nutrition. His whole system is fully given in a three volume book now placed in the library at the Nutrition Laboratory. He has a large number of publications on this matter. I found Von Pirquet a very excellent administrator, capable of getting nurses and hospital people to carry out his instructions very satisfactorily. The kitchens which I visited where food for "Hoover Feeding" was in preparation seemed most excellent. However, I could see no special advantage in the development of this new terminology. He thought it to be much more understandable by trained nurses and in the home of people.

Shortly before my visit in Vienna a group of German doctors who had been co-operating with the Quaker Child-feeding Committee in Berlin, visited Vienna to observe the work of Von Pirquet. I understand that they were well impressed with his work but not with his terminology and his system.

On the roof garden of the childrens clinic I saw a large number of children who were being treated for tuberculosis by the open air method. They lived practically without clothing, both summer and winter, sleeping as well as feeding and playing out of doors. They were as tanned as Indians and at first sight one would think they were of some colored race. The treatment appeared to be extremely successful. Von Pirquet thought it had something to do with the changed conditions in the blood due to the tanning of the skin. He did not say what part of the sun's light, whether the infra-red or the ultra-violet was the cause of the benifit from living in the sunlight.

At the Mauddling Home, I saw a large number of children reported to have rickets but certainly were in a very under nourished condition, some of them extremely small for the age which they were said to have. I never saw children have better medical attendance than these. The physicians seemed to be in excellent "rapport" with the patients, who were well fed and well attended to. They were using a large amount of cod liver oil in treatment of these cases. I spoke to them of the possibility of using some mixture of beef fat to replace the cod liver oil. I had previously learned from Dr. Mallenby in England (we spoke about the matter at the Physiological meeting in Paris) that he had made the suggestion that beef fat could be mixed with glucose and peanut oil and made palatable and would be the next best thing to cod liver oil to combat rickets. About three barrels of this mixture were made up and sent down to Vienna, I believe however they were in the hands of Dr. Chick who came from the Lister Institute and at the time of my visit in Vienna she had returned to London. I could find nothing about the result from the use of this glucose and beef fat mixture which, I believe, they called "Butol". In fact, I found the physicians of Vienna with whom I spoke not especially interested in it, because they had all the cod liver oil supplied to them that they wanted to use, in spite of the fact that the English looked upon the cod liver oil as being extremely expensive and wanting a substitute. At this Mauddling Home, it costs 40 cents a day to maintain a bed. They had about 200 beds. The bill for keeping there going was met by subscriptions from a group of people in New York City. They had plenty of space in which to enlarge this type of work and they had a tremendous waiting list of patients that should come in under their care, patients of practically the same class of need from under nourishment. No provision had been made for these nor was any in sight.

I had an opportunity for a short visit with Professor Paulta, formerly professor of pathology in the University of Vienna. He showed quite clearly the marks of under nutrition or at least, of weakness, which could not be charged up entirely to old age.

The streets of Vienna seemed rather empty, the city appearing too large for the population. There were many people very poorly clothed, some men bare footed and the clothing of the common professional man on close inspection showed that it was made of a combination from several suits and had probably been turned and changed one way or another more than once. I was informed on good authority that many of the middle class of professional people had found it necessary to sell things in their homes in order to provide food for their families. A competent young woman from the Laboratory of Dr. Cathcart was working in connection with the Friends Service Committee investigating the actual dietaries of a number of poor Austrian families these dietaries she was going to compare with data of similar nature gathered sometime ago in England. Such work as this will give the best indication of conditions in Vienna.

General Remarks

General Remarks.

Prior to 1914, almost each year found one or more responsible members of the Nutrition Laboratory staff visiting foreign institutes and laboratories in which work was being conducted in the same or closely related fields of research and frequently foreigners were in collaboration here. For example, in 1913, Dr. Benedict made a very extensive trip to foreign laboratories during the months, February to June. In the same year, Dr. Higgins was in Europe during the months from May to December. He collaborated in the study of the effects of alcohol on fatigue with Professor Galeotti at the Monte Rosa Laboratory. He visited also several other laboratories both in England and on the Continent and attended the Physiological Congress at Groningen. During the same year, at least three Europeans spent considerable time working at the Nutrition Laboratory: Dr. Carl Tigerstedt of Helsingfors, Dr. Hurschhauser of Dusseldorf and Dr. L. Bull of the Marey Institute, Paris. I came to the laboratory in the spring of 1914 when Dodge and Benedict were finishing their joint publication on the Psychological Effects of Alcohol (Carnegie publication 232) and very shortly after I arrived, Professor H. M. Smith went to Europe. He had scarcely more than reached Germany when the beginning of the war made it necessary for him to leave and return to America. This cordial cooperation and association with European laboratories has been interrupted for somewhat more than six years. I count it a very great privelege to have had the opportunity of representing our laboratory by going to Europe at this particular time. I cannot forbear expressing my heartiest appreciation to Dr. F. G. Benedict, Director, for making my trip possible and for so generously spending his time and thought in helping me with counsel and multitudinous arrangements. I know that much of the cordial welcome which I received was due to his previous wide acquaintance and long friendship with European workers. I wish also to express my thanks to my colleagues, Dr. T. M. Carpenter and Professor H. M.

Smith, for their willingness to assume extra burdens during my absence and to help me in many ways. I must mention also the faithful work of my assistant, Mr. E. S. Mills, during the months I was away.

My object in going to Europe was, therefore, to help to again establish a cordial relationship between foreign laboratories and ourselves; to familiarize ourselves with changes in personnel and organization and with the present activity of these laboratories as well as making our own work known to them. Personally it was my first opportunity to become directly acquainted with many of the workers and I was keen to discover techniques and methods in clinical physiology with adults which might be useful at the Nutrition Laboratory, and in general I considered it a unique opportunity to have some part in re-establishing international scientific relations at a time when the world has been so scourged with a spirit of hate.

Hardly any experience can be more stimulating to the scientific interest and zeal of a man than to come into personal contact with the workers in his own field. While it is possible to associate with these men at scientific meetings and congresses, much more in net results can be gained by visiting them personally in their own laboratories and institutions. Only such occasions allow for extended discussion of problems of mutual interest with apparatus and results at hand for observation and demonstration. Under these circumstances one is able to make quite an accurate judgment as to the man's methods of work and the reliability of his results and thus to more properly evaluate his publications, both past and future. Incidentally one discovers a good many useful details of laboratory equipment and management never shown in any way in publications.

As the previous pages show, I visited laboratories in England, France, Belgium, Holland, Denmark, Sweden, and to some extent in Germany and Austria. I attended three scientific meetings; The British Psychological Association, London; The British Physiological Association, Cambridge; and the Physiological Congress in Paris, giving papers at the two latter meetings. On June 16, in

GENERAL REMARKS

Amsterdam. I delivered an address on the work of the Nutrition Laboratory of the Carnegie Institution of Washington at the first public meeting in Holland regarding the establishment in that country of a National Institute for Nutrition. I was everywhere greeted with greatest cordiality, thus emphasizing in my own mind that the interchange of such visits is bound to result in a better international understanding.

Perhaps it does not sound well to say that on such a visit one makes it his partial aim to advertise the laboratory from which he comes. However, I believe it is the duty of a man employed by an institution such as the Carnegie Institution of Washington which was endowed with the purpose of "The improvement of mankind" to make every reasonable effort to familiarize the world with the work and scientific results produced by this laboratory.

Unfortunately, there is in Europe the impression that an American is always wealthy and can find money for any scientific project that may be mentioned, e. g. the publication of some big treatise or compilation. It seemed to me a duty to make clear to the European workers that American scientists have to be economical and that the war has greatly reduced the buying power of the incomes from all endowments left to scientific institutions; that therefore we do not at present merit the reputation which the earlier affluent days seems to have produced.

The unique importance of nutritional problems during the recent war with the persistent shortage of food still bitterly experienced in certain parts of Europe made it profitable for this laboratory to secure some first-hand data on present nutritional conditions in these countries, especially in the Central Powers, and to have a representative inspect to some extent the methods and work of relief organizations operating in the later countries. It is my conviction, after seeing the conditions, that the Americans would do well to render all possible assistance to these countries at present.

I came back with the opinion that experimentation in reference to the physiological effects of alcohol is still very important for the world in general. Europe is looking upon Prohibition in the United States as more or less of an experiment, and there probably is more interest in scientific data on alcohol experimentation than ever before.

When visiting laboratories in rapid succession, one is almost staggered by the multiplicity of workers and elaborate equipments and although many of the European laboratories are quite old, yet there are a number almost brand new and I was impressed that there are many things, both great and small, that the Americans can learn with profit from the Europeans.

263

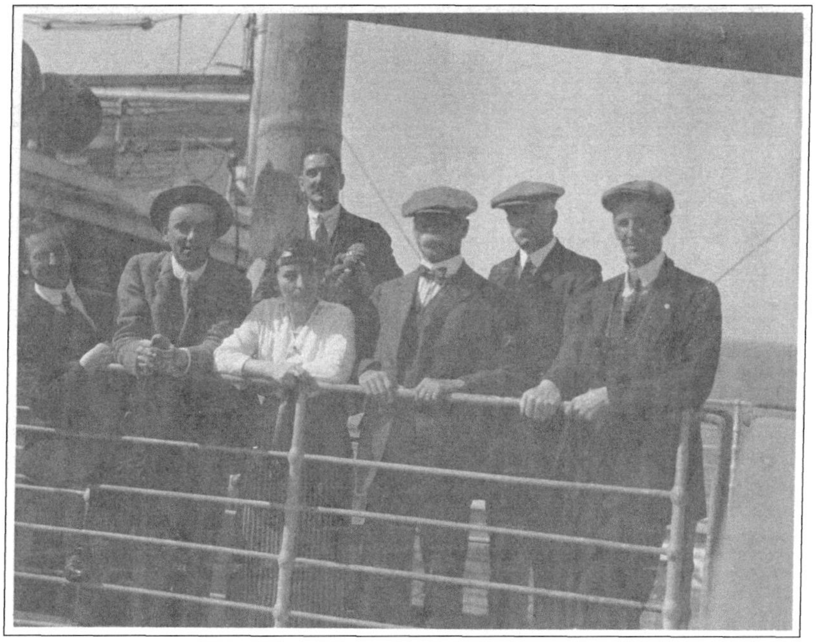

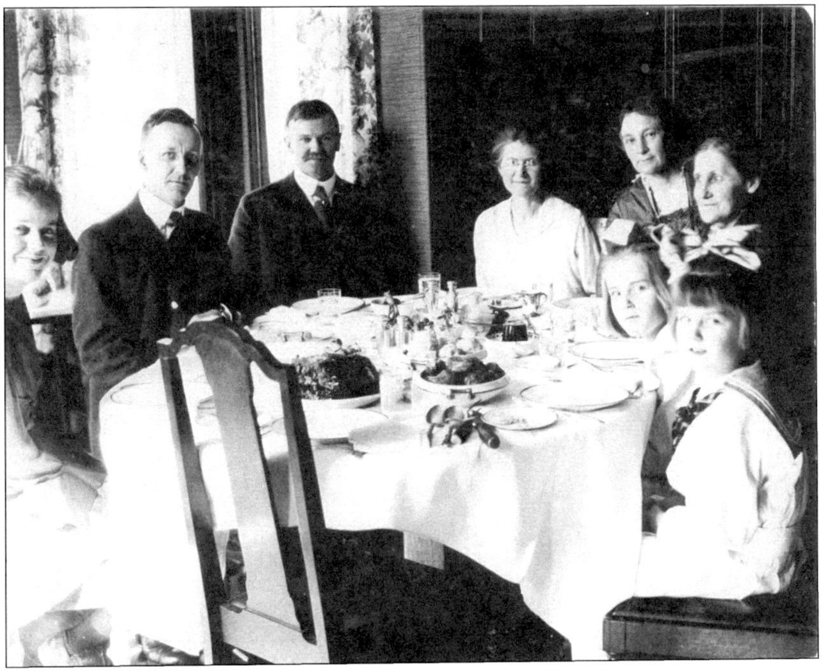

WALTER MILES AND HIS 1920 GRAND TOUR

GENERAL REMARKS

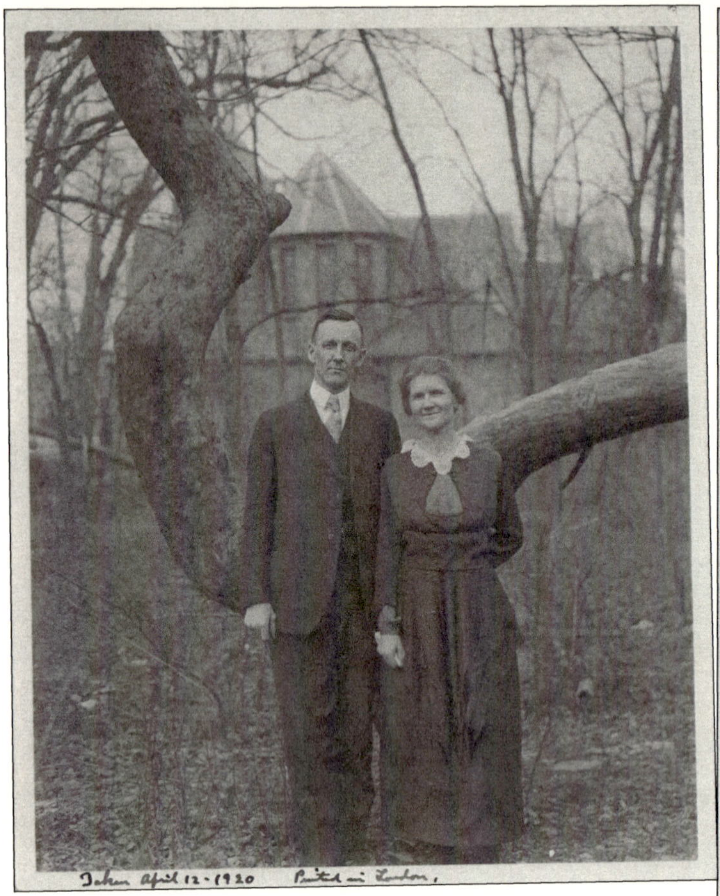
Taken April 12-1920 Printed in London,

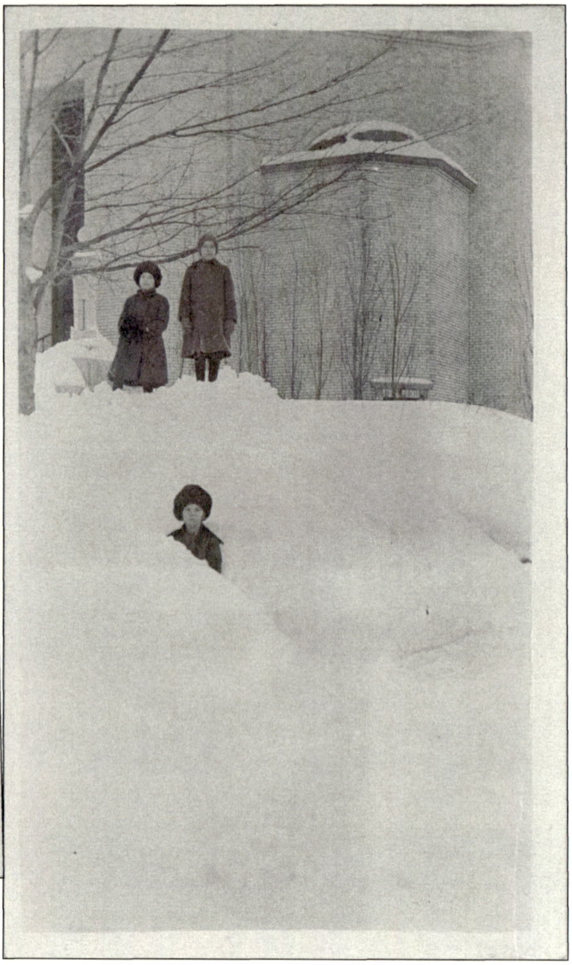

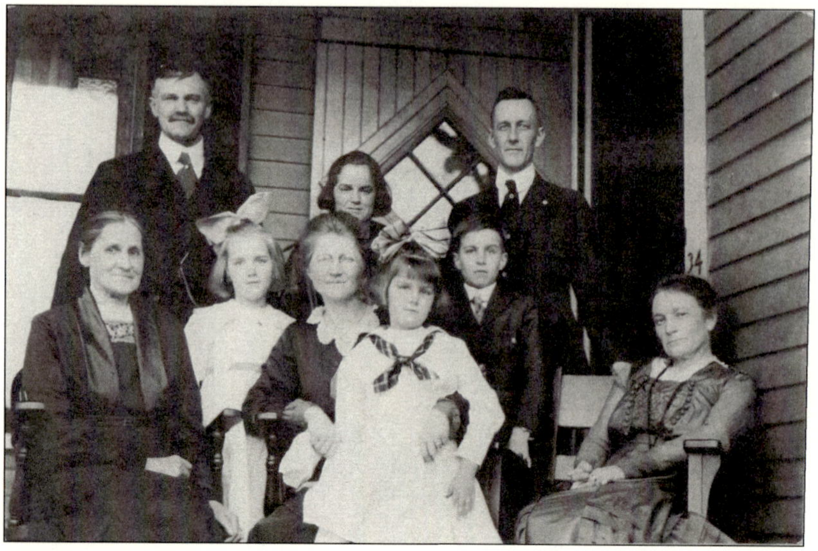

GENERAL REMARKS

265

Jan-Feb. 1920
G.H. du Paula Souza
of São Paulo, Brazil
was with us for
about 6 wks at
CNL.

Carpenter, Smith, Benedict, Souza, Miles

307

Hull, Mass. June 23-1920

GENERAL REMARKS

Received from Benedict June 1-1920

List of those to whom circular letter was probably sent regarding Professor Miles' visit to Europe.

- Alquier, Monsieur J., 16 Rue de l'Estrapade, (V^e), Paris, France.
- Amar, Dr. Jules M., 292 Rue Saint Martin, Paris, France. *Absent until June 15*
- Anrep, Dr. Gleb V., in Prof. Starling's Laboratory.
- Arrhenius, Prof. Svante, Nobelinstitut d. Akad. d. Wissensch., Stockholm, Sweden.
- Asher, Prof. L., Physiologisches Institut, Bern, Switzerland.
- Barcroft, Prof. Joseph, University, Cambridge, England.
- Bertrand, Prof. G., Pasteur Institute, 28 Rue Dutot, Paris, France.
- Cathcart, Prof. E. P., University, Glasgow, Scotland.
- Fano, Dr. G., Via Gino Capponi, n. 3, Florence, Italy.
- Gautier, M. Armand, Places des Vosges 9, Paris, France.
- Haldane, Dr. J. S., University, Oxford, England.
- Halliburton, Prof. W. D., Kings College, London, England.
- Hamburger, Prof. H. J., Institute Physiology, University, Groningen, Holland.
- Hasselbalch, Dr. K. A., Finsens medicinske Lysinstitut, Copenhagen, Denmark.
- Henriques, Prof. Valdemar, Physiol. Lab., University, Copenhagen, Denmark.
- Johansson, Prof. J. E., Karolinska Institutet, Stockholm, Sweden.
- Krogh, Dr. August, Ny Vestegade 11, Copenhagen, Denmark.
- Metzner, Prof. R., Physiological Institute, Basel, Switzerland.
- Pekelharing, Prof. C. A., Utrecht, Holland.
- Pembrey, Dr. M. S., Guy's Hospital Medical School, London, England.
- Philippson, Dr. M., Solvay Institute, Brussels, Belgium. *Away.*
- Rubner, Prof. Dr. Max, Physiol. Inst., University, Berlin, Germany. (Possibly sent?)
- Sir Edward Sharpey-Schafer, ~~Schafer, Prof. E. A.,~~ University, Edinburgh, Scotland.
- Sondén, Dr. Klas, Hotorget 14, Stockholm, Sweden.
- Tigerstedt, Prof. R., Helsingfors, Finland.
- Tigerstedt, Prof. Carl, Helsingfors, Finland.
- Waller, Dr. A. D., 32 Grove End Road, St. John's Wood, London, England.
- Wolf, Dr. C. G. L., School of Agriculture, Cambridge, England.

-2-

List of those to whom circular letter was probably sent regarding

Professor Miles' visit to Europe. (cont.)

    Bayliss, Dr. W. M., University College, London, England.

    Boissonas, Dr. Léon, Rue des Allemand 5, Geneva, Switzerland.

    Bull, Monsieur L., Marey Institute, Boulogne-sur-Seine, Paris, France.

    Douglas, Dr. C. Gordon, St. John's College, Oxford, Eng.

    Fredericq, Prof. Léon, University, Liège, Belgium.

    Hansen, Prof. C., Agricultural Expt. Station, Copenhagen, Denmark. *[vet. High School.]*

    Héger, Prof. Paul, Solvay Institut, 23 Rue des Drapiers, Brussels, Belgium,

    Hindhede, Dr. M., Frederiksberg Allé 28, Copenhagen, Denmark.

    Le Goff, Dr. Jean, 178 Faubourg St. Honoré (VIIIᵉ), Paris, France.

    Lindhard, Dr. J., University, Copenhagen, Denmark.

    Maignon, Dr. F., École Nationale Vétérinaire, Lyon, France.

    Richet, Prof. Ch., University, Paris, France.

    Santesson, Prof. C. G., Karolinska Institutet, Stockholm, Sweden.

    Sherrington, ~~Prof. C. S.~~ *Sir Charles S. Sherrington*, Dept. Physiol., Museum, Oxford, Eng.

    Slosse, Prof. A., Solvay Institute, Brussels, Belgium.

    Sorensen, Dr. S. P., Carlsberg Laboratory, Copenhagen, Denmark.

    Starling, Prof. E. H., University of London, London, Eng.

    Wiersma, Dr. E. D., Westerhade 10, Groningen, Holland.

    Zunz, Dr. Edgard, Solvay Institute, Brussels, Belgium.

    Zwaardemaker, Prof. H., Physiol. Lab., University, Utrecht, Holland.

Transcriptions

Page 1:
This is the carbon copy of the Foreign Report written for Nutrition Laboratory. It contains nearly all the pictures that are in that report and these have been numbered to correspond with the typewritten material. There are many other pictures not directly mentioned in the typed copy.

Page 10:
Figure 1. Institute of Physiology, located at Gower Street. This building faces on a court. The physiology work was mostly in the right hand end. The library and "Tea Room" were above the entrance. In the Tea Room a great collection of the photographs of physiologists trained here. Under the entrance is a tunnel way with doors opening into basement rooms and leading to the back where the animal cages are located. Physiol. Chem. is in the left hand side of the building.

Figure 2. Professor W. M. Bayliss at his laboratory desk in the Physiological Institute, University College, London.

Page 11:
Figure 4. Group of research students, Donegon, Anrep and Daly in Professor Starling's Laboratory. Note the overhead supply of power, gas and air. Anrep was trained with Pavlov.

Figure 3. Dr. I. de Burgh Daly experimenting with the string galvanometer and audion valves. This was an Edelmann galvanometer with a Cambridge string carrier, a combination which seemingly worked well.

Page 17:
Figure 5. "Tea Time" in the Psychological Laboratory, University College, London. Professor Fr. Aveling, Professor C. S. Spearman, and Capt. Philprit.

Photograph below Figure 5. Professor Spearman and Aveling. Taken at a guess on the focus, as I found the ground glass screen had been left in the library of the Physiology Institute. Went and recovered it before taking Figure 5 and never left it again. At Brussels I completely failed one picture of Professor Heger and myself by leaving in the slide of the screen focus camera—to my great regret.

Page 18:
Figure 6. Professor W. D. Halliburton in his office at King's College Medical School, London.

Page 34:
Figure 8. Sir George Thane, Chief Inspector for the Anti-Vivisection Act in England taking tea with Professor Pembrey. He said the newspapers must not see this picture.

Small photo below Figure 8. Dr. E. Mallenby of King's College for Women. Snap taken at Paris when at Physiological Congress.

Page 35:
Figure 7. Professor M. S. Pembrey in his office at the Physiological Laboratory at Guy's Hospital Medical School, London.

Page 41:
Figure 9. Professor A. D. Waller at his desk in the Physiological Laboratory, Imperial Institute, London, writing up some experiments on the physiology of plant growth.

Figure 13. Dr. and Mrs. Waller (and W.R.M.) in their garden studying the manner and rate of growth of the lupine.

Page 42:
Figure 12. Gas analysis in the Waller Garden, 32 Grove End Road, London. Miss de Decker determining the volume of respired air in one bag. Mrs. Waller writing the notes. Dr. Waller discussing how long to continue the walking experiment on the two young men shown in other views.

Figure 11. The pause for "tea" in the walking experiment. Seventy rounds of the garden had been completed, thirty more were to follow. (Left to right, Miss de Decker, Subject, Mrs. Waller, Subject, Peter, and Dr. Waller.)

Page 44:
Figure 10. Dr. and Mrs. Waller with their assistant Miss de Decker conducting an experiment on the CO_2 production during walking. Note the rubber tube for the subject to step on at each round. This registered by a pointer on the myograph.

Photograph below Figure 10. My exposure was too slow.

Page 57:
Figure 14. Dr. Leonard Hill and his assistant Miss D. Harwood-Ash working with the kata-thermometer at the National Institute for Medical Research, London.

Figure 15. Dr. H. H. Dale's Laboratory for Bio-chemistry and Pharmacology at the National Institute for Medical Research, London.

Page 63:
Figure 16. Dr. Edgar Schuster in his private machine shop at his residence, 110 Banbury Road, Oxford.

Figure 17. Dr. Schuster, Secretary of the Publication Department of the National Institute for Medical Research, London.

Page 64:
Figure 32. A circuit interrupter for use in investigation on nerve physiology. Made by Dr. Lucas but not described until 1921 by Adrian.

Page 69:
Figure 60. Prof. Schafer and family, host of 1923 Congress at Edinburgh. See opposite page.

Page 70:
Figure 19. Professor Schafer was convalescent and had not been out of doors for several days.

Figure 18. Professor Sir Edward Sharpey-Schafer at his residence, North Brunswick, Scotland.

Page 74:
Figure 21. Dr. J. C. Meakins at the Royal Infirmary working on blood samples.

Figure 22. The "Concertina" apparatus for continuous record of respiration. Used by Haldane, Meakins and Priestly. See Journ. Physiol. 1919, 52, p. 433.

Page 75:
Figure 20. Dr. W. W. Taylor in his laboratory for Physiological Chemistry, University of Edinburgh. The apparatus is for determining hydrogen ion concentration.

Page 78:
Figure 23. Professor A. Cushny in his newly equipped laboratory for Pharmacology at the University of Edinburgh.

Figure 24. Professor E. P. Cathcart looking over views from the Nutrition Laboratory at his desk in the Physiological Laboratory, Glasgow.

Page 79:
Figure 26. (Missing) Dr. J. Barcroft's large, three compartment respiration chamber made with glass walls. Mr. H. Secker, assistant, standing in middle compartment, in view below. Taken with camera in compartment A. The CO_2 absorber shows at the right of the bicycle ergometer.

Figure 25. (Missing) View taken from compartment C. One of the (F) food offerings shows in the lower left corner. Mr. Secker Barcroft's personal laboratory assistant in the view.

Page 90:
Figure 28. Dr. E. D. Adrian in his laboratory preparing a demonstration for the Physiological Society Meeting.

Figure 29. Various pieces of apparatus devised and used by the late Dr. Keith Lucas and now in the laboratory of Dr. Adrian.

Page 91:
Figure 30. The capillary electrometer and signal galvanometer devised by Dr. Lucas. Note the stone pillars and slabs on which the apparatus is supported.

Figure 31. The light from the electrometer passes through a small opening (x) to the camera which is in the dark room.

Page 106:
Figure 34. Professor C. S. Sherrington at his office desk, Physiological Laboratory, University of Oxford. He was working on some drawings of the semi-circular canals.

Figure 35. Professor Sherrington's laboratory for the course in advanced physiology.

Page 107:
Figure 37. Dr. H. C. Bazett (right) discussing his course in Clinical Physiology with W.R.M.

Figure 36. A new form of myograph devised by Professor Sherrington to take the place of the usual pendular style.

Page 112:
Figure 38. Dr. C. G. Douglas in his laboratory at the Physiology Department, University of Oxford. He was looking up a print in Carnegie Inst. Pub. 187.

Figure 39. A practical way to combine sampling tubes for using one mercury reservoir.

Page 113:
Figure 40. Dr. Douglas sitting inside his large respiration chamber. A reflection in the picture gives the appearance of an outside window at the back. I do not recall one but there is such in the chamber in Haldane's Laboratory.

Figure 41. The CO_2 absorber and ventilating system of the respiration chamber used by Dr. Douglas.

Page 116:
Figure 42. Dr. J. S. Haldane at his desk in his residence on Linton Road, Oxford.

Figure 43. The physiological laboratory in Dr. Haldane's residence. Dr. Haldane was assisted by his son, Dr. Jack Haldane (center) and by Dr. H. W. Davis of Australia.

Page 119:
Figure 44. Professor Georges Dreyer, Department of Pathology, University of Oxford. (It was a dark day and we were upstairs by a skylight. A slide rule, "my constant companion" he said, was in his hand.)

Figure 45. Mr. G. F. Hanson, an American, working with his newly developed gas-mixing meter.

Page 120:
Figure 46. Mr. H. F. Pierce with the splendid low-pressure chamber installation designed and built by himself.

Figure 47. Mr. Pierce and Mr. Hanson inside the chamber, indicating the commodious space for experimentation.

Page 131:
Figure 48. Dr. L. Bull making some adjustments on a large string galvanometer of his design at Marey Institute, Paris.

Figure 50. Details of apparatus used by Dr. Bull in drawing and silvering glass strings for the string galvanometer.

Page 132:
Figure 49. The Marey monument in front of the Institute. Dr. Bull holding the first motion picture camera.

Page 143:
Figure 51. Professor Louis Lapicque with apparatus for determining the "chronaxie" of nerve in the Physiological Laboratory, Sorbonne, Paris.

Figure 53. (Not Pictured) Dr. Stodel in his laboratory for Physiological Chemistry, Sorbonne, Paris.

Page 144:
Figure 56. (Not Pictured) The balcony above the court was a favorite promenade for Congress members. Here W.R.M. took some "snap-shots," Figures 57, 59, 60.

Figure 58. Prof. Langlois (left) demonstrating to Congress. I read my papers on the "Pursuitmotor" in this same session and room.

Figure 59. Prof. Richet (left) Pres. of Congress, discussing politics.

Page 148:
Figure 54. Professor Henri Pieron at his desk in the Psychological Laboratory, Sorbonne, Paris.

Page 148 (cont.):
Figure 55. Group at Societe Scientifique d'Hygiene Alimentaire in Alquier's apartment, Paris. Sitting: Mrs. Alquier and Alquier. Standing: Alquier (son), Bertrand, Lefevre, Lemoine, and Compton.

Page 172:
Figure 61. Professor A. K. M. Noyons at his office desk in the Physiological Laboratory, University of Louvain.

Figure 62. Professor Noyons (right) and Dr. Libbrecht (next) with assistants in their shop. The two round objects are shell cases being made into respiration chamber windows.

Page 173:
Figure 70. Partial view of the new Physiological Institute, University of Louvain.

Page 176:
Figure 71. The electrocardiogram outfit and room at the Physiological Institute, Louvain. A duplicate of Salomonson's apparatus.

Figure 72. View of electrocardiogram outfit showing arm-electrodes, and a second pillar with case for instrument. This is a corner basement room. Windows fitted with special black shades, walls white, the pillar of cement. Top has grove around the edge so that small things will not roll off. The case about the galvanometer made of several glass slides which can be taken out leaving only top and frame work.

Page 180:
Figure 67. Professor Noyons balancing electrical heat against animal heat in his large differential calorimeter.

Figure 69. Details of the differential calorimeter. A "Rotamesser" at the left. Mounted on and in the cabinet are the arrangements for regulating the water temperature in the two chambers.

Page 181:
Figure 68. The electrical control apparatus of the differential calorimeter.

Page 182:
Figure 63. Photograph of a chart showing the ground floor of Professor Noyon's differential calorimeter, Louvain.

Figure 64. Photograph of a chart showing the schematic side elevation of Noyon's Calorimeter. The view has been mounted upside down.

Page 183:
Figure 66. Photograph of a chart showing the arrangement for ventilating and regulating the temperature of Noyon's Calorimeter.

TRANSCRIPTIONS FOR PAGES 183–206

Figure 65. Photo of the chart showing the schematic arrangement of the water pipes and electrical system of Noyon's Calorimeter. Views upside down. This page turned wrong way in binding.

Page 194:
Postcard. (Not Pictured) This card given to me by Van Leersum. This is a view of Dr. Van Leersum's house, 28 Vondelstraat, Amsterdam. The house has a front of about 24 ft. and in style is just like the one to the left of it. Mrs. Van Leersum informed me that such a place cost 26,000 g. In the front part of the house, the basement is kitchen, 1st doctor office and hall, 2nd living room with balcony, 3rd chambers. The numbers are on the house walk, not on the door, and easy to see. The beam which extends out at the top is a regular feature of Amsterdam houses, and used to secure rope and pulley when moving and transferring heavy things in and out. Professor Wertheim Salomonson lives in a similar house a few down to the left in this photo.

Photograph. A photo of Dr. E. C. Van Leersum. He is a physician, medical historian and professor pharmacology, has traveled much and is a good linguist.

Page 195:
Figure 73. Dr. E. C. Van Leersum making an experimentation in his private laboratory at his residence, 28 Vondelstraat, Amsterdam.

Photograph below Figure 73. Mrs. Van Leersum died of heart disease Sept. 9, 1937, Leersum, Holland.

Dr. and Mrs. Van Leersum and their son-in-law Mr. Wopke de Gavere in the Leersum living room at 28 Vondelstraat, Amsterdam. This is the front room and on the second story, see view of the house on page 162. A small balcony opens off this room by a large door-window and we spent many happy times in this room. Dr. Van Leersum's office, two large rooms, was on the first floor. Mrs. (Wijsman) Van Leersum does wonderful work with her needle and always at it. Mr. de Gavere, a larger, pleasing man. The daughter Jacoba Gavere, had son Ate born May 1-1920, we all visited her and saw the fine boy.

Page 201:
Figure 74. Professor J. K. A. Wertheim-Salomonson in his electro-cardiographic laboratory, Neurological Institute, University of Amsterdam.

Page 202:
Figure 75. Professor Salomonson's electrocardiographic outfit of his own design.

Figure 76. Salomonson's small signal galvanometer at the left. Hill's (Lund) episcotister at the right.

Page 206:
Figure 77. Professor F. J. J. Buytendyk (standing) and his assistant Dr. M. N. J. Dirken on their rowing-machine ergometer.

Page 206 (cont.):

Figure 78. A calorimeter of the Atwater-Benedict type lately constructed by Professor Buytendyk, Amsterdam. Figs. – 79, 80, 81, 82, and 83 in the Nutrition Laboratory copy of this report are photos which were kindly given me by Professor Buytendyk.

Page 207
A photo of Buytendyk's calorimeter given to me by him.

Page 213:
Figure 86. Professor H. Zwaardemaker at his desk, University of Utrecht.

Figure 87. In Professor Zwaardemaker's office. Drs. Gryns, Zwaardemaker, and van Leersum.

Page 214:
Figure 84. Professor C. A. Pekelharing at his typewriter in his residence library, Utrecht. Dr. van Leersum at left.

Figure 85. Dr. van Leersum and Professor Pekelharing.

Page 218:
Figure 88. Professor W. Einthoven photographed by his original string galvanometer in his special laboratory for string galvanometer work, Leiden.

Figure 89. Group of scientific workers in Professor Einthoven's laboratory. Bytel, Einthoven, Prof. Verzar (Budapest), Dr. Liljestrand (Stockholm), Bergansius.

Page 220:
Figure 90. Second view of the group with Professor Einthoven June 18, 1920. A better likeness of Einthoven.

Page 225:
Figure 91. Professor W. Storm van Leeuwen at his special circular desk at his residence.

Page 231:
Figure 92. Professor H. J. Hamburger standing at his laboratory desk, Groningen.

Figure 94. Professor and Mrs. Hamburger in their garden. Praediniussingel 2, Groningen, Holland.

Page 232:
Figure 95. A memorial in tile given Professor Hamburger by citizens of Groningen and placed in the Physiological Institute after the IX International Physiological Congress.

Page 242:
Figure 96. Professor August Krogh in his laboratory studying the capillary circulation of blood, Copenhagen.

Figure 97. Respiration chamber and bicycle ergometer used by Dr. Krogh in studying muscular work on carbohydrate and fat diets.

Page 243:
Figure 98. Spirometer and a simple kymograph devised by Dr. Krogh. See the very small spirometer for study of insects.

Photograph below Figure 98. Professor August Krogh, photo taken 1918.

Page 253:
Figure 99. Drs. H. Mollgaard and A. C. Anderson photographed by the large ventilating pump of their respiration apparatus.

Figure 100. Recording apparatus connected with the Mollgaard-Anderson respiration chamber for cows, Copenhagen.

Page 254:
Figure 101. Dr. Erik M. P. Widmark in his laboratory with his distillation apparatus for determining the amount of alcohol in urine, blood or other fluid.

Page 264:
Figure 102. Professor J. E. Johansson at his desk, Karolinska Institute, Sweden.

Figure 103. General view in Johansson's laboratory for muscular work and respiration measurements. Prof. Johansson standing at the left, Mr. Yarl, mechanician at the ergometer.

Page 265:
Figure 104. Professor Johansson using the ergometer of his own design.

Figure 105. The Johansson ergometer is placed inside a respiration chamber special accommodation for which has been made in the floor of the laboratory.

Page 268:
Figure 106. A group at the residence of Professor Johansson 17a Kungsklippa St., Stockholm. Sitting: Sonden, Miles, Arrhenius, Johansson. Standing: Stenström, Strömbeck, Miss Johansson, Miss Geiges, N. Johansson.

Figure 107. A group in Professor Sonden's office in the new Technical High School, Stockholm. Sonden, Miles and Johansson. Portrait of Berzelius.

Page 307:
Top photograph. Jan–Feb 1920. G. H. du Paula Souza of San Paulo, Brazil was with us for about 6 wks. at CNL. (Below photograph) Carpenter, Smith, Benedict, Souza, Miles

Outline

London
- University College, Physiological Laboratory (Bayliss & Drummond)
    - "Most complete institute of its kind in London"
    - Bayliss—heavy teaching load—"war held large numbers of men from their medical courses, now these have returned"
    - During visit, Sir George Thane stopped by
        - "General supervisor under the Home Office for the Anti-Vivisection Act in England"
        - Previously a professor at University College
        - Discussion about regulations for animal research
        - Especially difficult to get permits for Pavlovian type of research
        - Bayliss hoped for permits for Anrep, who had studied with Pavlov (i.e., already knows the technique)
        - Strong antivivisection movement in England
    - Mention of Bose, but more on him later
    - Drummond—in class during WRM's visit; minimal contact
    - Anrep working in the physiology lab
        - Comments on Pavlov → last saw him March 1918, broken leg, work at a standstill, dogs dying (no food), sham feeding (gathering gastric juice)
        - Efforts underway to provide some international aid to Pavlov
        - Anrep working on digestion, work hampered by anti-vivisection forces
        - Ethical dilemma—knows of publishable work in Pavlov's lab, but doesn't want to publish it himself—it should go to Pavlov, but Pavlov cannot publish right now (no resources)
    - Daly—working with string galvanometer
        - Pilot during the war; has ideas about aptitudes of good pilots
        - Interested in WRM's photos of pursuit pendulum and pursuit meter
- University College, Psychological Laboratory (Spearman)
    - Arrived in time for tea—Spearman congenial host
    - Department cramped for space

- o Spearman—war-related research—"stereopsis in aviation"
  - ■ Accommodation of the eye at different distances
- o Also studying gun pointing ability and "quickness and keenness of observation"
- o Research on psycho-galvanic reflex by one of Spearman's students, "Miss Newmark"
  - ■ Stimuli—auto horn, falling chair, noxious tastes
- o WRM glad to hear of some degree of reconciliation between Spearman and Pearson ("at last made up their differences")
- o WRM tried to call on Pearson at the Galton Lab but was refused admission (didn't follow a rigidly proscribed set of procedures); WRM puts a good face on it, but clearly perplexed and a bit annoyed (letter from Pearson inserted, explaining procedures)
- University College Hospital, Heart Clinic (Lewis & Cotton)
  - o Examining men with "soldier's irritable heart"
  - o Room cold, and being touched by cold hands of examiners increased the nervousness of the men; highly detailed description of examination sequence
    - ■ Electrocardiograms taken with men in sitting positions
    - ■ Exam includes seeing how heart reacts to exercise (e.g., "ten hops on each foot'" "lift two 10 pound dumbbells" 30 times)
- Kings College Medical School, Physiological Laboratory (Halliburton)
  - o WRM complimented Halliburton on *Physiological Abstracts*
  - o Like Bayliss, complained about student overcrowding as a result of men returning from the war
  - o Mentioned efforts to get funds to Pavlov (funds raised, not yet delivered)
  - o Learns news of psychologist C. S. Myers (left Cambridge, now in London practicing psychiatry)
- Guy's Hospital Medical School, Physiological Laboratory (Pembrey & Ryffel)
  - o Pembrey—rugged constitution; research on exercise and effects on heart
  - o Against birth control—"each family should have more children and let the best survive"
  - o No sympathy for those with "nervous breakdowns"
  - o Traditional view of male-female roles in society (concerned about current disproportion of females) (war effect)
  - o Alcohol research—studying "relation of drink habit to infant mortality"
  - o Anatomy museum—best wax models (including some of skin diseases) ever seen by WRM
  - o Photo with Thane—he visited in his capacity as inspector for government Anti-Vivisection agency
    - ■ Insisted photo taken by WRM not be shown to press (shouldn't be seen taking tea with a physiologist)
- London University, Imperial Institute, Physiological Laboratory (Waller)
  - o Plant physiology work—in conflict with Bose (who claims plants have

physiological systems that are essentially nervous systems); dispute is "loud and hot"
- Research on psycho-galvanic reflex
  - WRM as demo subject—electrodes on palm and back of hand and on forearm
    - Stimulus—leg shock—strong response on hand, none on forearm (normal); some Ss respond on forearm and Waller thinks "they are those individuals who are mediumistic and double personalities" (thus, psycho-galvanic response as a classification tool)
    - Waller not very familiar with (extensive) U.S. research on the topic
- Research on muscular exercise and carbon dioxide (stair climbing and walking around the Waller estate)—method "rough and ready" and not very precise

- Lecture at University of London Club (Bose)
  - SRO
  - Bose proposed to show that "plants respond to stimulat[ion], that they (certain of them) sleep, that they have connective tissue which corresponds to the nerves of animal organism, that these plants show action currents as do the nerves of animals, . . ."
  - Tried to demonstrate a plant's "death spasm"
  - WRM sees connection between Bose's ideas and his Hindu religion
- Kings College for Women (Mallenby)
  - Research on rickets in dogs
  - Alcohol research, $N = 1$, large amount of alcohol, had S copy drawings (decrease in accuracy with increased consumption)
- Ministry of Health, Whitehall, London (Newman)
  - Sir George Newman—chief medical inspector of England
    - Spoke of Sherrington and Starling as the best of current British crop of physiologists
  - Criticizes L. Hill (next stop for WRM)
    - Too much time on social action and not enough time on science
  - Discussed alcohol and its adverse effects
- National Institute for Medical Research, Mount Vernon (Hill & Moore)
  - Interested in the cooling power of air to aid work
  - I/O—interested in "fitness for work of the employees under different factory conditions"
  - Strong advocate for the benefits (on health) of "the outdoor life" (mentions effect of open air on TB and argues that open air increases metabolism)
  - Not appreciated in the scientific world, he claimed
  - With Brownlee—large mechanical calculator
  - Schuster—first "Galton Fellow"; worked with McDougall, "arranged a modification of McDougall's Dotting-Machine" (hand-eye coordination apparatus)

- o Discusses the difficulties in preparing manuscripts for publication
- British Psychological Association, Bedford College, Regents Park (Scripture & Klein)
  - o E. W. Scripture talk on "Speech inscriptions in normal and abnormal conditions" (main point—voice inflection signals "psychological states or conditions")
    - Also argued that "voice curves" were a good way of diagnosing epilepsy
    - Scripture now "practicing medicine" in London
  - o Klein gave a talk on camouflage, relevant for discussions of "vision and space perception"

Edinburgh

- University of Edinburgh, Physiological Laboratory (Schafer & Taylor)
  - o Problems with German scientists employed there—asked to resign at the outset of war (one did, one didn't)
  - o Schafer thinks men more creative as scientists than women
- University of Edinburgh, Royal Infirmary (Meakins)
  - o Developing method of treating pneumonia with oxygen
- University of Edinburgh, Laboratory of Pharmocology (Cushny)
  - o Includes museum of "natural drugs, poison arrows, etc." ("curare")
  - o Discussed work of "Liquor Control Board" in England
    - WRM had a discussion, at lunch with Meakins, of how a man's eyes change when drunk
- University of Edinburgh, Psychological Laboratory (Drever)
  - o Drever, background in education, recently taken over lab after death of "Dr. Smith"
    - Interested in social psychology and at odds with McDougall
  - o Best equipped psychology lab of any that WRM sees in Great Britain
    - "Everything that Spindler and Hoyer have made"
  - o Background—lab founded by George Combe (phrenologist)
  - o Lab currently can get anything it wants "by asking the trustees of the Combe estate"

Glasgow

- University of Glasgow, Physiological Laboratory (Paton & Cathcart)
  - o Cathcart had been a research associate at Carnegie Nutrition Lab
    - Another comment about the overflow of students, not able to attend during WWI
    - Interested in psychic research; and in the value of psychology for turning average men into fighting soldiers
    - Cathcart praised the research from the Carnegie Nutrition Lab, but complained that because the reports were so long and so detailed,

they had less effect than they should have ("no one or very few read them"); failed to feature the key points and findings
- University of Glasgow, Psychological Laboratory (Watt)
    o Currently converting an old house to a psychology lab (no more room in the physiology lab)
    o Educated in Germany, knew Marbe when in Külpe's lab, British citizen, POW in the war
    o In Wundt's lab for a time—volunteered as a "subject" but was not accepted
        ■ Thought it was because those in W's lab "were very choice in the matter of whom they used as a subject and did not want to get in anyone who might have introspections at variance with the views of the professor"
        ■ Believed it was "suicidal for an outside student to contradict or debate the point of view of the man in charge in a German laboratory, that is, as a usual thing. He found Professor Külpe was of a very different kind and gave his students much freedom."
    o Did his research with Külpe before WWI and just before the war, returned to visit; war broke out and he was imprisoned for two years ("very trying experience both mentally and physically"); released in exchange for a German soldier in prison in England
    o Interested in psychology of sound; in some kind of disagreement with Titchener

Cambridge

- University of Cambridge, Physiological Laboratory (Barcroft, Adrian, Hartree)
    o Large chamber for studying varying levels of oxygen
    o Barcroft recently spent six consecutive days and nights in the chamber
        ■ Bicycle ergometer used to see how reduced oxygen would affect exercise
            □ "Martin Boyle ergometer"
        ■ Did some memory experiments with Bartlett while there (no results reported)
        ■ Last 24 hours—Barcroft "quite miserable"
    o Adrian studies nerve and muscle physiology
        ■ Much ado about getting access to "stop-cock grease" for "gas analysis apparatus"
    o Hartree—physicist; galvanometer design
- University of Cambridge, Psychological Laboratory (Bartlett, Muscio)
    o C. S. Myers has left the department and gone to London to practice psychiatry
        ■ "Much success during the war with shock cases"
    o Psychology lab one wing of the new Physiology Lab; separate budget; designed by Myers
        ■ Less apparatus than at Edinburgh

- - - Kraepelin ergometer; Rivers-McDougall dotting machine; apparatus for psychogalvanic reflex
    - Good audio equipment, including records of "primitive music" from "a wide range of tribes" (Rivers anthropological work)
  - Bartlett—currently working on "non-experimental material"
    - WRM describes the War of the Ghosts study without realizing it is a memory study
      - Describes it as an "anthropological problem involving accuracy of report"
    - "A large, quiet man"
  - Muscio—been working for British Industrial Fatigue Board
    - Looking at time of day and fatigue in factory work
    - Apparatus—"Match Board" (sounds similar to Purdue pegboard)
    - Suggested WRM go to London to visit Fatigue Board, in part to "find more concerning Myers plans for the National Institute for Applied Physiology and Psychology" (WRM unable to go though)
- University of Cambridge, Biochemical Laboratory (Hopkins, Peters, Grey)
  - Hopkins—work on vitamins and effects on rats
  - Peters—on paramecium
  - Grey—biochemistry
- Meeting, British Physiological Society (monthly)
  - WRM presentation—"Two Measures for Muscle Coordination"
    - Described pursuit pendulum and a "test for static control" (steadiness test with the ataxiometer)
      - Good reception—friendly questions from Bayliss and Halliburton
  - Met for first time A. V. Hill and W. H. N. Rivers (discussed alcohol research with latter)
  - On trip back to London, discussion of "psychical research" and alcohol effects with several psychologists and physiologists (question raised about the viability of placebo controls in alcohol research)
  - WRM impressed with the meeting, in part because of its monthly arrangement (cuts down on socializing because they all see each other regularly—more on science)
- Cambridge and Paul Instrument Company (Whipple)
  - Whipple especially interested in skin temperature work
  - WRM sees latest improvements in string galvanometer

Oxford

- University of Oxford, Physiological Laboratory (Sherrington, Bazett, Douglass)
  - Sherrington—same problem mentioned earlier—overcrowding of students in postwar, resulting in difficulty in completing research
    - Lab has the largest myograph (measures force generated by muscle) WRM has ever seen

- - -
    - Sherrington "keen on the subject of alcohol research"; likes test of static control
    - WRM impressed with Sherrington's "genuineness and kindly eagerness"; "the very antithesis of self assertion and so ready to point out things of interest in your work rather than discussing the importance of his own"
  - Bazett—working on vital capacity and physiology of breathing; effects of exercise (raising dumbbells) on heart rate and bp; the "Flack Blowing Test"; work on decapitated animals
    - Worked on "problems with the personnel of aviation" during the war
  - Douglas—muscle exercise (bicycle ergometer) and perspiration
    - Large respiration chamber
- Haldane's private laboratory (Haldane)
  - Has a respiration chamber similar to the Douglas's
  - Just finished a study like Barcroft's—living in the chamber and decreasing oxygen levels
    - "Subject was unable to write and he was very stubborn and hard to deal with"
- Pathological Laboratory (Dreyer)
  - Really doing physiology, not pathology
  - Effects of weight, sitting height, and vital capacity on metabolism
    - Diagram of device for measuring sitting height (highly reliable method)
  - Work with tuberculosis—could identify patient progress better with vital capacity data than by doctor in "personal contact" with patients
  - High praise for Carnegie Nutrition Lab from Dreyer
  - Worked on problems of aviation during the war (ability to withstand high altitude effects)
    - Goats as Ss
- University of Oxford, Psychological Department (McDougall)
  - No apparent laboratory or apparatus (perhaps due to "present crowding by medical students")
  - McDougall recently in Zurich with Jung
  - Alcohol effects
    - McDougall thought moderate amount relieved tension and allowed one to "proceed with his work with much greater comfort"
    - Thought this neglected by those "over-enthusiastic for the prohibition of alcohol"
    - McDougall about to come to Harvard as Department Head in the fall

Paris

- Marey Institute (Bull, Nogues)
  - Bull—during war, using reaction time as a way to estimate distances from artillery and locate enemy guns; Bull adapted a string galvanometer for the

same purpose—six string galvanometers, each with microphone, each at a different distance from the gun; problems in the field—couldn't distinguish sound waves from gun firing from sound waves of shell traveling through space; when problem solved by a British scientist ("Tucker Microphone") the device was widely used—Germans knew of it and had batteries at different distances fire simultaneously to confuse the device, but the system continued to work
- Marey generates considerable income from the production and sale of string galvanometers
- Also devised device to photograph bullet as it leaves a gun
- Much detail on the process of "making and silvering glass fibres for the string galvanometer"
- WRM impressed with stereoscopic photos that Bull made of apparatus when he visited Carnegie Nutrition Lab ("I think I never saw any pictures which more accurately portrayed apparatus")
  - Nogues—work on high speed motion pictures of humans
- Physiological Laboratory of the Sorbonne (Lapique, Stodel)
  - Lapique—war-related nutrition research
    - Using sea moss and seaweed as food for animals (chickens)
    - Also interested in nerve physiology (threshold stimulation)
    - Depressed—had difficulty securing his post after the death of a colleague—trouble getting apparatus
    - At time of visit, busy (with Stodel, a chemist) preparing to host upcoming Congress of Physiology
- Psychological Laboratory of the Sorbonne (Pieron)
  - Pieron (with wife) investigating "sensitivity of different portions of the retina to different qualities of light stimulation"
  - American student in residence—David Weschler
    - Student of Woodworth's at Columbia
    - Trying to develop a measure of what we would call today emotional intelligence (Miles skeptical)
  - WRM "attracted to Pieron"—successor to Binet, only does research; possibility of organizing an "Institute of Psychology" in Paris to include Pieron, Janet, and Dumas
- Pasteur Institute (Bertrand)
  - Bertrand asked if WRM drank alcohol (no response) and shared McDougall's view that in small amounts, alcohol had the effect of "freeing one from inhibition and making one more comfortable which would at times outweigh in importance any small decrease in one's production efficiency"; thought lab results only a part of the story; astounded at Prohibition in U.S. (no comment from Miles)
    - B studying problems of safely canning vegetables

- WRM—Bertrand "an example of the very highest type of scientific Frenchman"; dismayed that Bertrand been blocked from the National Academy
- Societe Scientifique d'Hygiene Alimentaire (Alquier, Lefevre)
  - Alquier—building respiration calorimeter (using plans of Lefevre); hopes to secure help of Carnegie Nutrition Lab
  - No real lab at present; WRM suggested they get a "portable Benedict respiration apparatus" and get started modestly on some research
  - WRM gently suggested that the calorimeter that desired might be outdated and easily replaced with simpler apparatus
  - Alquier appears angered—"The honor of France" demanded the best apparatus possible (but Lefevre seemed to agree with Miles)
    - WRM reports some criticism (by Lapique) of the institute during the war, concerning its insistence that only white bread was nutritious
    - WRM's general impression is that the institute has accomplished little if anything
- Physiological Congress of 1920 (July 16–20)
  - WRM completed most of his visit prior to this conference but reports on it here because he has just been describing the French labs
  - "Germans and Austrians excluded" from the conference
  - German physiologists held a meeting in Hamburg in June, inviting physiologists from neutral countries; some went (e.g., Van Leeuwen, Einthoven)
  - Germans looked on Paris meeting as something pushed through by the French and "another indication of the French attitude to crush out the Germans"
  - Germans felt betrayed by Professor Rubner
    - Rubner had apparently put his authority behind government reports that Germany had plenty of food, when he really knew otherwise
  - Dutch want the next conference, after Paris, to be in America
    - Hinted that Americans supply a ship to transport European physiologists
    - Think Americans all wealthy
  - Congress attendance → N = 401 on program but standard procedure is to send in an abstract to get on the program
    - Real attendance about 300 (mostly French and British); 8 (plus Miles) from U.S.
  - Important presentation and demo by Noyons (Louvain) of a "differential calorimeter" used for dogs and other small animals
  - WRM gave presentation on the pursuit meter—comments on the lack of satisfaction over giving or listening to conference presentations
    - Problem of concurrent sessions
    - Translation issues
    - Good to see people, but even better to see them in their labs

- o Americans resistant to the idea of hosting the next conference
  - ▪ Think few would attend given the "depreciation in the value of European money, which may continue for some years to come"
- o When in Germany, WRM had sounded out German physiologists about the future of international meetings
  - ▪ Germans happy to meet with Americans and maybe the British "but with the French, never!"
- o Next Congress proposed (and accepted) to be in Edinburgh in 1923

Brussels
- Solvay Institute, Physiological Laboratory (Heger, Phillipson)
  - o Heger—university about to be relocated and he was heavily involved in the planning
    - ▪ First rate physiology lab, although Heger no longer involved (in administration)
    - ▪ Phillipson, then on vacation, now in charge of the lab
    - ▪ Comments on war and German occupation—Heger bitter toward German scientists "carried on such a propaganda of absolute misrepresentation with the objective of breaking the moral of the Belgian people that he never again could really believe statements made by these scientists"

Louvain
- Catholic University, Physiological Laboratory (Noyons)
  - o Noyons had been assistant to Zwaardemaker at Utrecht for 10 years
  - o Went to Louvain in 1912 ready to design and build a new physiology lab, then war broke out
  - o Noyons Dutch and therefore a neutral, even though he was living in Belgium
  - o Despite trying conditions (Germans kept trying to burn down his house), able to finish much of the construction in wartime
  - o Shows WRM the differential calorimeter that he would take to the Paris conference (above)
  - o Has some apparatus for olfactory and auditory sensation
  - o Great detail in explanation of N's string galvanometer for electrocardiogram records and respiration chamber

Amsterdam
- National Institute for Nutrition (Van Leersum)
  - o Van Leersum—noted historian of medicine
  - o The institute was established by Van Leersum after University of Leiden failed to develop a lab for him
    - ▪ Toured North American labs in the process of planning the institute

- - - Was at Carnegie Nutrition Lab for about two weeks (and so knew WRM)
  - - Arranged for WRM to give a talk on the work of the Carnegie Nutrition Lab
      - First public meeting of the new Institute
    - Van Leersum wishes to obtain apparatus from Carnegie Nutrition Lab—especially the Benedict portable respiration apparatus
    - "His attitude and the way he is going at the problem in Holland seems to be in marked contrast to the method of the French"
      - Refer back to Societe Scientifique d'Hygiene Alimentaire
- University of Amsterdam, Neurological Institute and Laboratory (Salomonson)
  - Salomonson quite talented with equipment
    - Seems a combo of engineer, physicist, and physiologist
    - Nicely designed string galvanometer
    - Has arranged for it to be manufactured by the Cambridge and Paul Scientific Instrument Company (Cambridge, England)
  - Studying sensitivity of nerves and muscles to electric shock
    - WRM couldn't follow the math
- Free University (Amsterdam), Department of Biology (Buytendyck, Dirken)
  - Established by religious organization, so no Darwin taught there and vivisection not allowed
  - Buytendyck—studies learning in animals, respiration physiology, and work physiology
  - Respiration work—rowing machine adapted to an ergometer (photo)
  - Roughly built calorimeter in the cellar of the building
  - Student of Buytendyck had an offer to go to West Virginia University (in U.S.) and study in psychology department of a "Professor Morse"; WRM had no useful info, which surprised the student

Utrecht

- State University of Utrecht, Institute of Physiology (Pekelharing, Zwaardemaker, Ringer)
  - Accompanied by Van Leersum
  - Pekelharing—retired but still writing
  - Zwaardemaker—must be 60 but seems much younger
    - Studying olfaction, and sound waves generated by voice
  - Ringer—working in what used to be Donders' lab
- State University of Utrecht, Institute of Inorganic Chemistry (Cohen)
  - Chem Lab—not much of interest

Leiden

- State University of Leiden, Physiological Laboratory (Einthoven)
  - Einthoven—originator of string galvanometer

- Research focus is on theoretical problems related to the instrument
- Lab specially designed to have galvanometers free from any vibration (solid brick pillars)
- Commented on the hope to return to truly international meetings
- WRM noted that in groups of scientists, E tended to gravitate toward the younger men, unlike most senior scientists
- In the war, sympathetic to the Allied side, but reticent at the Paris meeting
  - State University of Leiden, Pharmaco-Therapeutic Institute (Van Leeuwen)
    - Van Leeuwen had visited Carnegie Nutrition Lab and met WRM
    - After visit, afflicted with "American disease"—depressed about the quality of labs (well financed) compared to those in U.S.
  - State University of Leiden, Physical Laboratory (Onnes, Crommelin)
    - Lab designs thermometers
    - WRM—"At such visits one devoutly wishes that he knew more so he could understand and ask more but the days of for Aristotelian knowledge have passed for most men"

Groningen
- State University of Groningen, Physiological Laboratory (Hamburger)
  - Similar to the work of Bazett in Oxford
  - No change in this lab from Benedict's 1913 description
  - 1913 Congress had been held here (demonstrations by Starling)
    - Through Hamburger's efforts at the time, a trolley line was run out to the university (now called the "Hamburger Line")
- State University of Groningen, Psychological Laboratory (Heymans, Brugmans)
  - Visited in the company of Hamburger
  - Labs seem brand new
  - Heymans & Brugmans doing visual perception research
    - Illusions, depth (stereoscopes)
  - Experiment comparing color naming speed with speed of reading color names underway
  - Experiment on thought transference underway
    - Experimenter thinks of a square on a checkerboard-type array
    - Subject who claimed to have telepathy then tries to put his finger on the square
      - Claim—more correct answers than chance would predict
      - But researchers felt the procedure still needed to be tightened up

Copenhagen
- University of Copenhagen, Animal Physiological Laboratory (Krogh)
  - With his wife (Dr. Marie Krogh), just finishing a manuscript on "the relative efficiency for muscular work of carbohydrate and fat diets"

- Used respiration chamber and ergometer for the study
            - Carbohydrates more efficient
        - Planning for research on the effect of "alcohol as food for work"
            - Spoke highly of Widmark (Lund)
        - Disagrees with L. Hill (London) on the idea that metabolism is higher when measured outdoors
        - Sees no point in the French desire to build a calorimeter
        - Studying capillary circulation of blood
            - Interested in subject of skin temperature and how it relates to capillary circulation
        - Has a gas analysis apparatus he hopes to use in mines to detect high levels of harmful gasses
        - Krogh "is the perfect embodiment of the quiet, humble but keen, indefatigable scientific worker"
        - Expertise—the physiology of respiration
    - University of Copenhagen, Physiological Laboratory (Henriques)
        - New electrocardiogram the only addition since Benedict's 1913 visit
            - Galvanometer probably better than the one manufactured by Cambridge Instrument Company in England
            - Studying changes in electrocardiogram with changes in position of the subject
    - University of Copenhagen, Psychological Laboratory (Lehmann)
        - Studying mental effort and metabolism
        - Missed Lehmann (on vacation); knows Dodge very impressed with Lehmann's work
        - Research similar to that reported by Carpenter (of Carnegie Nutrition Lab) on a visit in 1911
    - Veterinary High School (Mollgaard, Anderson)
        - Studying respiration physiology with farm animals, especially cows
            - Carbon dioxide produced in intestine extracted via fistula
    - Dansk Maalerfabrik (Gjellerup)
        - "Bohr wet meters" used extensively in respiration work manufactured here
            - Cost $125, four times the price before the war
            - WRM seemed to question this, but was told costs for materials have also quadrupled

Lund

- Karolinska University, Biochemistry Laboratory (Widmark)
    - Some controversy over whether Widmark will get a permanent position or whether it will go to a German professor
        - Newspapers involved, concerned about a German invasion of such professors
    - Lab old and inadequate

- WRM interested in Widmarks's research on measuring the content of alcohol in blood and urine
    - Widmark—after a dose of alcohol, concentration in blood and urine is the same
- Karolinska University, Physiological Institute (Thunberg, Westerland)
    - Thunberg includes statistics as part of the physiology course; notes importance of considering "things in terms of averages but also in relation to their variations from average"

Stockholm

- Karolinska University, Physiological Laboratory (Johansson)
    - Elaborate respiration chamber with ergometer
    - Good equipment but not being used to capacity
- Stockholm Board of Health & Technical High School (Sondén)
    - Nothing of note
- Military Hospital (Stenström)
    - Stenström had visited Carnegie Nutrition Lab so Miles knew him
    - Research on respiration physiology
    - Recent interest in electrocardiology

Hamburg

- University of Hamburg, Physiological Laboratory (Kestner)
    - Stopped on way from Holland to Denmark
    - Lab established just prior to war, no significant equipment
    - German physiologists met here in June (prior to Paris meeting where they were unwelcome)
    - Kestner depressed
        - Lab not likely to be equipped any time soon
    - Two dogs prepared "in the manner of Pavlov for the production of gastric juice"
    - At scientific meetings, Kestner thought Germans would meet with the Americans and the British, but not the French
        - "He sited (sic) the situation at Frankfurt and the French use of colored troops."

Berlin

- University of Berlin, Physiological Laboratory (Rubner, Gildemeister)
    - In Berlin for five days, going from Sweden to the Paris meeting
    - Able to pass through Germany because of Quaker connections (U.S. had not at that point made official peace with Germany)
    - Rubner "distant and formal"
        - Hard to do research at present because of "the problem of how to secure more food"

- ■ Expressed no animosity to U.S., just a bit toward England, "but quite the reverse in reference to the French"
- o Lab—no new equipment since before war
- University of Berlin, Institute for "Arbeitsphysiologie" (Thomas)
  - o New institute—Thomas in charge
  - o Physiology of work the main problem, but at present all efforts on problems of food, especially "utilization of old potatoes"
  - o Inflation effects—no money for journals
  - o While in Berlin (and Hamburg), looked at work by "American Friends Service Committee" concerning feeding undernourished German children in schools
    - ■ Developing an efficient system

Leipzig

- University of Leipzig, Institute for Experimental Psychology (Kruger, Kirschmann)
  - o Kirschmann (native German) had been at Toronto, but returned to Germany in 1913–14 to recover from mild stroke; stuck when war broke out
  - o Dismissed from Toronto
  - o Taken on as assistant by Wundt (same position he held as a student of W in 1892)
  - o Wundt now 88, retired, and in the country (note: Wundt died on 31 August 1920, shortly after the Miles visit to Leipzig)
    - ■ Kruger now the director
  - o Leipzig lab "most severe" (not even the original lab)
    - ■ No comfort; narrow, crowded rooms
  - o Kirschmann doing vision research
    - ■ Developed a stereoscope with images that could be tilted—tested the limits of stereoscopic vision
    - ■ Claimed to have been the first to "record the nature of the eye movements during reading"
- University of Leipzig, Physiological Institute (Garten)
  - o Institute famous through the "directorship of Ludwig and Hering"
  - o Garten—felt "cut off from the scientific world"
    - ■ During war, engaged in research on aviation and pilot selection
      - □ Developed a simulator that WRM tried
    - ■ Interested in string galvanometer and electro-physiology
      - □ Lab is full working order and busy

Vienna

- University of Vienna, University Hospital, Kinderklinik (Von Piquet, Schick)
  - o Wished to see Exner, but he is retired and living in the country
  - o Von Pirquet—testing undernourishment in children

- Hunger in daily life evident just "passing down a street"
- Not enough food, so need a system for selecting those most undernourished
  - "Pelidisi method"
    - Combines sitting height and weight
  - Sacratama method"
    - Measures "blood fatness, skin fullness, and muscles"
  - Children retested every two months
  - Nourishment precisely quantified
    - Nem = nourishment value of 1g of mother's milk
    - Nem = nutrition, element, milk
- "Hoover feeding" system in place
- Roof of children's clinic—WRM sees TB children being treated with "open air method"
  - Highly tanned, which Von Pirquet thought was the key to the success of the method
- Maudling Home—saw children with rickets
  - Treatment—cod liver oil
  - WRM suggests beef fat, "mixed with glucose and peanut oil
    - On suggestion of Mallenby (England); cod liver oil considered expensive in England
    - Viennese not interested
- Streets of Vienna mostly empty

General Remarks
- 1st visit from Carnegie Nutrition Lab personnel since before the war
- Goals
  - Reestablish relationships of long standing
  - Familiarize with present activity of the labs and pass along what Carnegie Nutrition Lab has been doing
  - For Miles, to become directly acquainted with other researchers
  - Discover techniques that could be useful at Carnegie Nutrition Lab
- Visiting labs better than just seeing scientists at conferences
  - More in depth understanding of their research and apparatus
- Another aim—promote and advertise Carnegie Nutrition Lab
  - But Miles had to correct the impression that Carnegie Nutrition Lab was well off financially
- Another aim: "persistent shortage of food" in postwar Europe provided opportunity "to secure some first-hand data on present nutritional conditions in those countries, especially in the Central Powers"
  - WRM noted the strong need for assistance
- WRM came back convinced of importance of alcohol research—Europe is looking to U.S. Prohibition as an experiment

Biographical Profiles

Adrian, Edgar Douglas (1889–1975). E. D. Adrian was a British physiologist and physician who conducted his clinical work at St. Bartholomew's Hospital in London and earned his MD in 1915. Adrian acted as a lecturer at Trinity College and later succeeded Joseph Barcroft as professor of physiology at the University of Cambridge. He remained at Cambridge until his retirement in 1951. Adrian studied the nervous system, including the electrical impulses caused by stimuli which caused pain, as well as olfaction and the electrical activity of the brain. He shared the 1923 Nobel Prize in Physiology or Medicine with Charles Sherrington for his work regarding the functions of the neuron.

Barcroft, Joseph (1872–1947). Joseph Barcroft, an Irish physiologist, earned his BA from the University of Cambridge in 1896. In 1900 he accepted the position of lecturer in natural science, which he held until he succeeded J. N. Langley as professor of physiology. He is best known for his research regarding the respiratory function of blood, which he studied during two high altitude expeditions that he also led—Teneriffe (1910) and Monta Rosa (1911). He invented a differential blood gas manometer and his classic book, *The Respiratory Function of the Blood,* was published in 1914. He delivered the Dunham lectures at Harvard Medical School in 1929, which were eventually published as *Features in the Architecture of Physiological Functions* (1934). Barcroft spent the latter part of his career studying the physiology of the developing fetus, and was knighted in 1935.

Bartlett, Frederick Charles (1886–1969). F. C. Bartlett was a noted British psychologist, famous for his work on cognition, especially his research on memory and on thinking. His graduate training was at Cambridge, where he completed the memory studies that were eventually published in his best-known work, *Remembering: A Study in Experimental and Social Psychology* (1932). In 1922 he became director of the psychology laboratory at Cambridge, remaining in the position until 1952, and training a generation of British research psychologists, including Donald Broadbent, who was later famous for his research on selective attention. Bartlett was knighted in 1948; three years later he was awarded the Royal Medal, Britain's highest scientific honor.

Bazett, H. C. (1885–1950). H. C. Bazett was a British physician who earned his MD from Oxford in 1910. He was awarded the Radcliffe Traveling Fellowship in 1912. Bazett spent twelve months at Harvard University doing postgraduate research before returning to Oxford to work as a demonstrator in pathology. He lectured at Oxford in clinical physiology from 1919 to 1921, when he accepted a professorship in physiology at Pennsylvania State University. Bazett remained at Penn State until his death in 1950. He studied temperature control and blood circulation and often used himself as a subject, attaching different thermocouples to his own arm, which were then attached to galvanometers.

Bertrand, Gabriel (1867–1962). Gabriel Bertrand was a French chemist who is best known for his discovery of the element manganese in plant life and his subsequent development of the concept of trace elements. He earned his PhD from the École de Pharmacie in Paris in 1904 and worked as an assistant to Albert Arnauld from 1890 to 1900. Bertrand was hired by the Pasteur Institute in 1900 as a staff member of the Biochemistry Institute and eventually chaired the department from 1908 until his retirement in 1937.

Bose, Acharya J. C. (1858–1937). A. J. C. Bose, a native of India, earned a BA from Cambridge and a BSc from London University. He then returned to India, becoming a professor of physics at Presidency College, Calcutta. There he developed a controversial theory that plants have nervous systems analogous to those of humans and could therefore respond to electrical stimulation and feel pain. In 1917 he created the Bose Institute in West Bengal, India, a pioneering research laboratory of biophysics, botany, and biochemistry that continues its work today. Bose frequently lectured in England, and in 1920 became the first native of India to be made a Fellow of the Royal Society. The British government also knighted him.

Bull, Lucien (1876–1972). Lucien Bull was an amateur photographer and an assistant to Étienne Jules Marey. He joined the Marey Institute in 1895, developing and printing chronophotographic negatives. Following Marey's death in 1904, Bull began his own research, which included high-speed studies of insect flight. He later became the subdirector of the institute. Bull hosted Raymond Dodge for four months during 1909–1910 and built much of his own apparatus. During World War I he constructed high-speed photographic analyses of ballistics for the British Army and conducted research in acoustics, physiology, and optics.

Buytendijk, Frederik Jacobus Johannes (1887–1974). F. J. J. Buytendijk, a student of Hendrik Zwaardemaker, earned his MD from the University of Amsterdam in 1909 and spent several years thereafter working in the labs of C. S. Sherrington, J. L. Langley,

A. V. Hill, M. Verworn and T. W. Englemann. In 1914 he was appointed lecturer of biology and physiology at the Protestant Free University in Amsterdam and in 1919 was promoted to professor. In 1925 Buytendijk accepted a professorship in physiology at Groningen University and around this time became interested in biologically oriented philosophy and animal psychology. He accepted a position in 1946 as professor of general and theoretical psychology at the University of Utrecht and eventually headed the psychology department. He was involved in the existential-phenomenological movement in psychology and issues in education and mental health. His important publications include *On Pain* (1943), *General Theory of Human Posture and Movement* (1948), and *Woman* (1950).

Cathcart, Edward Provan (1877–1954). E. P. Cathcart, a Scottish physician, earned his MB (1900) and MD (1904) from Glasgow University. He studied bacteriology and chemical pathology at laboratories in Munich and Berlin and worked in the Lister Institute of Preventative Medicine from 1902 to 1904. In 1905, Cathcart returned to Glasgow University to accept a lecturer position in physiological chemistry, which he held until 1915. While at Glasgow, Cathcart took two leaves of absence—in 1908 to spend five months in Ivan Pavlov's St. Petersburg Laboratory studying animal operative techniques, and in 1912 to spend a year with F. G. Benedict at the Carnegie Nutrition Lab researching nutrition and energy metabolism. Following his time in Boston, he published his classic monograph *Muscular Work, A Metabolic Study with Special Reference to the Efficiency of the Human Body as a Machine* (1913). In 1919 he accepted the position of Gardiner Chair of Physiological Chemistry at Glasgow and in 1928 succeeded Nöel Paton as the Regius Chair of Physiology, a position he held until his retirement in 1947.

Cohen, Ernst Julius (1869–1944). Ernst Julius Cohen was a Dutch chemist who received his PhD from the University of Amsterdam in 1893. He remained in Amsterdam and spent the following nine years working as a demonstrator and conducting research in the laboratory of J. H. van 't Hoff, including his well-known 'tin research' which refers to the allotropic modifications of tin. In 1901 Cohen became professor of chemistry at Amsterdam and in 1902 accepted a position as professor of inorganic and general chemistry at the University of Utrecht. In 1904 he oversaw the opening of the van 't Hoff Laboratory at Utrecht and he worked as director of the lab and professor of chemistry until his retirement in 1939. In 1922 he helped organize an international meeting of chemists in Utrecht and in 1925 took a leave of absence to spend a year at Cornell University as the George Fisher Baker Non-Residential Lecturer. On March 3, 1944, Cohen was transported to the Auschwitz concentration camp, where he was eventually killed.

Douglas, Claude Gordon (1882–1963). C. G. Douglas was a British physician who earned his BSc in 1906 under J. S. Haldane and his BM in 1908, both from Oxford University. In 1907 he was elected a fellow and lecturer at St. John's College, Oxford and held the position until his retirement in 1949. He continued to collaborate with Haldane on the problem of respiratory physiology. In 1911 Douglas traveled with Haldane, Yandell Henderson, and E. C. Schneider to Pike's Peak in order to study breathing at high altitudes, a research problem investigated by J. Barcroft the previous year. Douglas invented the Douglas Bag, a bag used for collecting expired air in order to determine oxygen consumption, for use on the Pike's Peak expedition. In 1924 he collaborated with J. G. Priestly to publish *Human Physiology: A Practical Course*. In 1942 Douglas accepted a position as professor of general metabolism at Oxford and remained there until his retirement.

Drever, James (1873–1950). James Drever was a Scottish educator and psychologist. He earned an MA in the liberal arts from the University of Edinburgh, taught in secondary schools for about 10 years, and then earned a BSc in psychology from London University in 1909. He continued as an educator and in 1918 was named the Combe lecturer in psychology at the University of Edinburgh. The laboratory at Edinburgh was exceptionally well-equipped because of a bequest from George Combe, the famous Scottish phrenologist. As director of the laboratory, Drever established a reputation as an outstanding teacher of experimental psychology. In 1931, the university established its first chair of psychology and Drever was named to the position, making him the first professor of psychology at a Scottish university. Although he published little, his 1916 book *Instinct in Man* was considered an improvement on McDougall's theory of instincts.

Dreyer, Georges (1873–1934). Georges Dreyer was born in Shanghai, the son of a Danish naval officer. He earned his MD from the University of Copenhagen in 1900. Dreyer spent the next several years traveling and conducting research in various laboratories on topics such as bacteriology, the diphtheria toxin, and the agglutination of bacteria. In 1907 Dreyer accepted a position as chair of pathology at Oxford University. During World War I he worked to properly diagnosis typhoid and paratyphoid, and suggested that all troops be inoculated with the typhoid-paratyphoid vaccine, a procedure that was eventually adopted. During the war he also devised an apparatus that allowed for the automatic supply of oxygen to pilots at different altitudes. Dreyer also devised a method for diagnosing syphilis, and is perhaps best known for his work regarding the blood volume of mammals and the importance of an animal's weight in regards to the lethality of toxins.

Einthoven, Willem (1860–1927). Willem Einthoven was a professor of physiology at the University of Leiden in the Netherlands, and also a physician, a general practitioner.

He was best known for inventing the string galvanometer, a device that measured the electrical activity of the heart and significantly advanced the measurement of cardiac activity beyond an earlier device invented by Augustus Waller. The string galvanometer was the predecessor of the modern electrocardiogram. The device could also be adapted to measure any rhythmic activity with electrical characteristics (e.g., nervous system function). His research on cardiac function earned Einthoven the 1924 Nobel Prize in Physiology or Medicine.

Grey, Egerton Charles (1887–1928). E. C. Grey completed his medical studies at Guy's Hospital in London and earned his DSc from Cambridge University in 1920. He researched the chemical action of bacteria at the Lister Institute and conducted biochemical research at the Pasteur Institute and the Biochemical Laboratory at Cambridge University. Following the end of World War I, Grey accepted an appointment as professor of chemistry at the University of Cairo, where he published *Practical Chemistry by Micro-Methods* (1925). He left Cairo in 1926 in order to conduct research for the League of Nations on nutrition and food in Japan, a project which lasted six months. He returned to Cambridge and began research on the chemical action of B. coli communis and the subsequent publication was in press at the time of his death in 1928.

Haldane, John S. (1860–1936). J. S. Haldane was a Scottish physiologist known for his research concerning the effects of various gasses on human physiology. He earned an MD from the University of Edinburgh in 1884. He taught at the University of Glasgow, New College, Oxford, and the University of Birmingham. During World War I, in the aftermath of German poison gas attacks, Haldane went to the war zone and attempted to determine the composition of the gasses being used. As a result of this work he invented an early version of a gas mask. Haldane also investigated the effects of dangerous gasses (especially carbon monoxide) in coalmines on the health and safety in miners, and in the 1890s he introduced the idea of placing small animals (e.g., mice, canaries) in mines as an early warning system. His son became a famous evolutionary biologist, known for playing a key role in what became known as the modern evolutionary synthesis.

Halliburton, William Dobinson (1860–1931). William Halliburton was a physician and physiologist, earning an MD from University College London in 1884. He held the chair of physiology at King's College London for 34 years. Halliburton was among the first to recognize the importance of chemistry for medicine and is considered to be one of the founders of modern biochemistry. He completed important research on the nature of proteins in the blood and in muscle tissue. He was a Fellow of the Royal Society (1891) and the Royal College of Physicians (1892).

Heymans, Gerardus (1857–1930). Gerardus Heymans was a Dutch philosopher and psychologist, and a professor at the University of Groningen from 1890 to 1927. He established a laboratory there in 1892, the first in the Netherlands. Heymans was known for his investigations of visual illusion, especially the Zöllner illusion, and his adaptation of Galtonian questionnaires for the study of various mental and emotional phenomena and the study of gender differences. With a student, Henri Brugmans, Heymans completed a series of controlled studies of mental telepathy (described by Miles in his report), claiming better than chance performance by some subjects.

Hill, Leonard (1862–1952). Leonard Hill was a noted British physiologist, trained in medicine at University College London. From the turn of the century until 1914, he taught physiology at the London Hospital Medical College. In 1914 he was appointed director of the department of applied physiology at the National Institute of Medical Research. He was known for his research on "caisson disease" (decompression sickness or "the bends"), and the effects of air circulation on body cooling. He was a strong advocate of the importance of the "cooling power" of air circulation on the efficiency of both mental and physical work, famously recommending that the ventilation and air movement in the British House of Commons be improved, with the aim of enhancing the quality of debate. Hill also argued for ventilation improvements in factories to enhance productivity. He was elected a Fellow of the Royal Society in 1900 and knighted in upon his retirement in 1930.

Hopkins, Frederick Gowland (1861–1947). F. G. Hopkins was a British chemist who began his work as an assistant to Thomas Stevenson at Guy's Hospital in London. He later earned his MB from Guy's Hospital in 1894, after working for several years without a formal degree. In 1898 Hopkins accepted a lecturer position in chemical physiology at Cambridge University and in 1914 was elected chair of biochemistry. He was awarded the 1929 Nobel Prize in Physiology or Medicine for his research on growth-stimulating vitamins, an award he shared with Christiaan Eijkman of Utrecht University. Hopkins was knighted in 1925 and served as the president of the British Royal Society from 1930 to 1935.

Johansson, Johan E. (1862–1938). J. E. Johansson was a Swedish physiologist and professor of physiology at the Karolinska Institute in Stockholm. In 1890 he traveled to Alfred Nobel's lab in Sevran, France and conducted research on blood transfusions. From 1904 to 1926, he worked as a member of the Nobel Committee for Physiology or Medicine. He is known for his research on pulmonary tuberculosis in Stockholm and its frequency in relation to population density and economic status.

Kestner, Otto (1873–1953). Otto Kestner was born Otto Cohnheim in Poland and received his MD in 1896 after studying in Leipzig and Heidelberg. In 1897 he accepted a lecturer position in the Physiological Institute at Heidelberg and began the research which would become his greatest discovery—the detection of the enzyme erepsin which he found was responsible for breaking peptones down to free amino acids. In 1913 Cohnheim accepted a position as full professor at the Institute of Physiology at Hamburg and at this time changed his name to Otto Kestner, his mother's family name. In 1934 Kestner was dismissed from Hamburg by the Nazis and fled to Britain. He worked briefly within the school of agriculture at Cambridge University and returned to his post in Hamburg following the end of World War II.

Kirschmann, August (1860–1932). August Kirschmann was a German experimental psychologist who earned a PhD under Wilhelm Wundt at the University of Leipzig in 1892. Starting in 1893, he taught and directed the laboratory at the University of Toronto. At Toronto he was known for his research on the perception of color; a color mixer he developed became a standard piece of laboratory apparatus. On a leave of absence in 1908 Kirschmann toured European laboratories, but fell ill and never returned to Toronto. He remained affiliated with the university until 1915, when the Canadian university officially severed ties. Kirschmann always believed the Toronto firing was due to the war and because he was German. In 1917 Wundt offered Kirschmann a position as laboratory assistant at Leipzig and Kirschmann remained there until his retirement in 1931.

Krogh, August (1874–1949). August Krogh earned his MS in zoology (1899) and his PhD (1903) from the University of Copenhagen. He accepted a lecturer position at Copenhagen in 1908 and was promoted to professor in 1916. Krogh married Marie Jörgensen in 1905 and they collaborated often throughout their careers, including a 1908 trip to Greenland to study the nutrition of Eskimos. They installed a respiration chamber and conducted metabolic studies, and upon returning to Copenhagen began their well-known work on the exchange of gases in the lungs. Krogh was awarded the 1920 Nobel Prize in Physiology or Medicine and later published the results of his award-winning research as *The Anatomy and Physiology of the Capillaries* (1922).

Krueger, Felix Emil (1874–1948). F. E. Krueger (Kruger in the Miles report) was a German experimental psychologist who earned a PhD under Wundt at the University of Leipzig in 1898. His research area was phonetics and auditory perception. Krueger held professorships at several German universities (Kiel, Halle, Leipzig), a Brazilian university, and spent a year (1912–1913) as an exchange professor at Columbia University in

New York. In 1917 he succeeded Wundt as director of the Leipzig laboratory. In 1935 he was elevated to the rector of the university, but was forced to retire for criticizing government policy toward Jewish professors.

Lapicque, Louis (1866–1952). Louis Lapicque (Lapique in the Miles report) was a French physiologist who spent his career as a professor of physiology and director of the laboratory of general physiology at the University of Paris, Sorbonne. His wife, Marcelle (de Herdia) Lapicque wrote her 1903 dissertation on the nerve impulse as a student of her husband, and they collaborated in the laboratory throughout their careers. Together they coined the term chronaxie in reference to the electrical excitation of the nerve and the study of it as a wave form over time. Lapicque was captured by the Germans during the occupation of France and spent three months in prison, during which he wrote *La machine nerveuse* (1943). Louis Lapicque died in 1952 and Marcelle Lapicque assumed his responsibilities as director of the laboratory until her death in 1962.

Mallenby, Edward (1884–1955). Edward Mellanby studied physiology at Cambridge and was a lecturer in physiology at King's College for Women in London from 1913 to 1920. In 1920 he became professor of pharmacology at the University of Sheffield and was chair of the department from 1920 to 1933. He is known for his investigations of rickets, a bone disease resulting in impaired growth, weakness, and bone pain. With dogs as subjects, he determined that a dietary deficiency produced rickets and the problem could be alleviated by repeated doses of cod-liver oil. Although he did not realize it, his research set the stage for the discovery that a lack of vitamin D was the cause of rickets (cod liver oil has a high concentration of the vitamin).

Muscio, Bernard (1887–1926). Bernard Muscio was a native of Australia who trained in both philosophy (with James Ward) and psychology (with C. S. Myers) at Cambridge. He returned to Australia and taught for a time at the University of Sydney, developing an expertise in industrial psychology. Muscio is known for his work on the effects of fatigue on work productivity. In 1919 he returned to Cambridge and began work with the British Industrial Fatigue Research Board, completing several experimental investigations of fatigue in British factories. In 1922 he returned to Sydney, where he remained on the university faculty until his early death from heart disease. He founded and was the first president of the Australian Association of Psychology and Philosophy.

Myers, Charles S. (1873–1946). C. S. Myers was a prominent British psychologist who co-founded the British Psychological Society in 1901. As a student, he joined William McDougal and William H. S. Rivers in a pioneering anthropological expedition to the Torres Straits. He earned his MD (1901) and ScD (1909) degrees from Cambridge. While a faculty member at Cambridge (1904–1922), he established the laboratory of experimental psychology (1913). Myers and Rivers co-edited the *British Journal of Psychology* for many years. During World War I, Myers was given a commission in the Royal Medical Corps. In a 1915 article in *Lancet*, he coined the term shell shock, considering it a psychological disorder (and not simply cowardice, the view held by the military at the time). During the 1910s and 1920s, his *Textbook of Experimental Psychology* was the leading textbook in England for training students in basic laboratory psychology. After World War I, his interests shifted to applied psychology and he became a leading industrial psychologist, co-founding the National Institute of Industrial Psychology in 1921.

Pearson, Karl (1857–1936). Karl Pearson was a British mathematician known as the founder of modern statistics because of his contributions to the field of statistical analysis. He is best known as the creator of Pearson's r for calculating the strength of correlations and for a version of the chi-square test. He was also an accomplished historian, an expert in German literature, a student of physics and physiology, and student of the law. He was a devoted follower of Francis Galton and a firm believer in eugenics. After Galton's death in 1911, the Galton Chair of Eugenics was created at the University of London. Pearson was the first person named to the position. Pearson also created and led the department of applied statistics at the University of London. He wrote a three-volume biography of Galton and was a co-founder of the journal *Biometrika*.

Pekelharing, Cornelis Adrianus (1848–1922). C. A. Pekelharing was a Dutch physician who began practicing medicine in 1872 after studying at the University of Leiden. In addition to his medical practice, he also served as a laboratory assistant to the physiologist A. Heynsius at the University of Leiden. In 1877 he accepted a position at the state school of veterinary medicine in Utrecht and in 1881 was promoted to professor of pathology and pathological anatomy at the University of Utrecht. In 1888 he was appointed professor of physiological chemistry and histology and conducted research on anthrax, blood coagulation, and protein digestion. Pekelharing also conducted laboratory research on the effects of alcoholism and was interested in improving the diets of the poor and undernourished. He helped establish the Netherlands Institute for Nutrition in Amsterdam and E. C. Leersum was the Institute's first director.

Pembrey, Marcus Seymour (1868–1934). M. S. Pembrey was a British physiologist and author of several well-known medical textbooks. As a medical student at Oxford (MD, 1895) he served for a time as a demonstrator in physiology for Burton-Sanderson and a research collaborator with J. S. Haldane. In 1900 he joined Guy's Hospital in London as lecturer in physiology and remained there until his retirement in 1933. He was much in demand as a speaker, known for inciting debate with his strongly held traditional views, especially concerning women, marriage, and the family. For example, he believed a women's sole responsibility was to marry as early in life as she could and have as many children as possible while she could (he and his wife had 10 children).

Peters, Rudolph Albert (1889–1982). R. A. Peters was a British physician who studied physiology and biochemistry at Cambridge University and obtained his MD from St. Bartholomew's Hospital in 1915. While at Cambridge he studied under Joseph Barcroft and J. N. Langley and conducted research with A. V. Hill. He worked under Hopkins in the biochemistry laboratory at Cambridge and in 1923 accepted the Whitley Chair of Biochemistry at Oxford University. While at Oxford, he conducted research on vitamin B and nutrition. Peters retired from Oxford in 1954 and accepted a position as head of the biochemistry department at the Agricultural Research Center. There he researched the toxicity of certain compounds in nature, specifically those plants that threatened livestock. He retired from the Agricultural Research Center in 1959 and returned to Cambridge where he remained active in scientific research until 1976. He received the Royal medal in 1949 and was knighted in 1952.

Pieron, Henri (1881–1964). Henri Pieron is considered the one of the founders of experimental psychology in France. He was trained in both philosophy and physiology at the University of Paris and was best known for his research into the physiology of sleep. Following the death of Alfred Binet in 1911, he was appointed director of the psychology laboratory at the Sorbonne in Paris and held that position during the time of the Miles visit. With Rene Legendre, Pieron completed a study in which they deprived dogs of sleep, and then extracted a substance they called "hypnotoxin" from the blood of the dogs. When injected into normal dogs, the substance induced sleep. He also became involved in applied psychology, developing expertise in the area of vocational guidance.

Rivers, William H. R. (1864–1922). W. H. R. Rivers was trained as a physician, but was better known for his contributions to anthropology and psychology. In 1901, he co-founded the British Psychological Society with C. S. Myers; the two also co-edited the *British Journal of Psychology* for many years. As an anthropologist, he teamed with two of his students, William McDougal and C. S. Myers, in leading a pioneering anthropological expedition to the Torres Straits. During World War I he applied psychotherapy procedures to treat shell-shocked British officers at Craiglockhart War Hospital, near Edinburgh, Scotland, believing shell shock to be a psychological disorder, not simply a matter of cowardice (the prevailing view). His work at Craiglockhart was featured in the 1997 film, *Behind the Lines*.

Salomonson, J. K. A. Wertheim (1864–1922). J. K. A. Wertheim Salomonson was a Dutch physician who also studied physics and neurology during his time at the University of Leiden. Upon completing his studies he worked as an assistant to Pieter Pel at the clinic for internal diseases at the University of Amsterdam and also worked in private practice. In 1893 Salomonson accepted a head position at the Municipal Clinic for Nervous Disease and Electrotherapy and left in 1900 to accept the position of chair in neurology and röntgenology at the University of Amsterdam. He collaborated with Ernst Cohen to create a Röntgen tube and within several months they were able to exhibit photographs of patients they had examined using X-rays. He is known for his research regarding the physiology and the nervous system and his adaptation of the Zeiss demonstration ophthalmoscope for photographic purposes.

Scripture, Edward Wheeler (1864–1945). E. W. Scripture earned a PhD in experimental psychology under Wundt at the University of Leipzig in 1891. He developed a first-rate psychology laboratory at Yale University, where he taught from 1892 to 1902 and also established an in-house journal, *Studies from the Yale Psychology Laboratory*. While at Yale, Scripture wrote two books describing the "new" laboratory study of psychology, *Thinking, Feeling, Doing* (1895) and *The New Psychology* (1897). After leaving Yale, Scripture went to Europe, earning an MD from the University of Munich in 1906. He remained in Europe, becoming known for his research in phonetics and speech disorders.

Sherrington, Charles S. (1857–1952). C. S. Sherrington was the Holt Professor of Physiology at Liverpool from 1895 to 1913, and then the Waynflete Professor of Physiology at Oxford until his retirement in 1936. He was known for research on reflex action and studies establishing the existence of synapses in neural circuits. His best-known work was *The Integrative Action of the Nervous System* (1906), based on a series of lectures given in 1904 at Yale University. His central argument was that the nervous system operates as an integrated whole, rather than as a set of separate processes. Sherrington was elected a Fellow of the British Royal Society in 1893 and was awarded the Royal Medal in 1905 and the Copley Medal in 1927. He won the 1932 Nobel Prize for Physiology or Medicine. During World War I he was too old for military service, but he chaired the Industrial Fatigue Board, and at one point worked long hours (13-hour shifts) at a shell factory in Birmingham.

Spearman, Charles (1863–1945). Charles Spearman was a British psychologist famous for his research and theorizing on intelligence and his work in statistics. He came to psychology relatively late after spending fifteen years in the military. In 1906, at the age of 43, he earned a PhD in psychology under Wundt at the University of Leipzig, and joined the faculty at University College London, remaining until he retired in 1931. He published papers that made him a pioneer in the use of factor analysis and he developed a nonparametric correlation procedure for ordinal data, known as Spearman's rho. He is best known for developing a two-factor theory of intelligence, which proposed that any behavior requiring intelligence involved some combination of a largely-inherited general intelligence factor ("g") and one of more specific factors ("s") related to the task at hand. Spearman was a follower of Galton and supported eugenic principles.

Thane, George Dancer (1850–1930). G. D. Thane was a student and then professor of anatomy at University College London, eventually chairing the anatomy department from 1877 until 1919, when he retired. He was knighted in 1919. At a time when physiology was a more prominent science at University College, Thane argued with some success that the study of anatomy should have equal status. He was known for the precision of his anatomical descriptions, many of which appeared in standard texts. In the later years of his career and continuing into his retirement, Thane became an inspector of laboratories for the government, checking for compliance with Britain's Antivivisection Act.

van Leeuwen, Willem Storm (1882–1933). Willem Storm van Leeuwen worked as an assistant to Rudolf Magnus at the Pharmacological Institute in Utrecht from 1911 to 1914. He earned his MD in 1912 and acted as deputy director of the Institute from 1915 to 1917. In 1919, he traveled to the United States in order to tour laboratories and hospitals. Upon returning to Holland he accepted a position as professor of pharmacology at the University of Leiden and conducted his well-known research on allergies, specifically asthma. He constructed an allergen-free room at Leiden and through his work discovered that asthma was the result of an allergy—usually resulting from the inhalation of mold or dust.

von Pirquet, Clemens (1874–1929). Clemens von Pirquet was an Austrian physician who received his MD from the University of Graz in 1900. He worked as a clinical assistant to Theodor Escherich until 1907, studying immunity and hypersensitivity, and eventually coined the term allergy. In 1908, he accepted a position as professor in the newly established department of pediatrics at Johns Hopkins University, a position he held for a year before returning to Vienna. In 1911, he succeeded Escherich as the director of the Santa Anna Children's Hospital and University Kinderklinik. As director, he researched and promoted prevention and recognition of infectious diseases, and stressed the importance of the bedside clinical observation.

Waller, Augustus Desire (1856–1922). A. D. Waller was born in Paris and in his adolescence his family moved to Scotland. He was educated at the University of Edinburgh, earning an MD in 1881. Waller became a lecturer in physiology at St. Mary's Hospital in London in that same year, and remained affiliated with the hospital for the rest of his life. He also taught physiology at the University of London and worked as a consultant at the National Heart Hospital. Waller is best known as a pioneer in the study of cardiac function and the inventor of the first functioning device for measuring the electrical activity of the heart. He was the first to use the term "electrocardiogram," and gave frequent demonstrations of his device using his dog as a subject. His version of the electrocardiogram was eventually replaced by a more precise apparatus, the string galvanometer, invented by Willem Einthoven.

Watt, Henry Jackson (1879–1925). H. J. Watt was a Scottish experimental psychologist at the University of Glasgow. As a graduate student, he was a member of Oswald Külpe's famous Würzburg school, earning a PhD on the experimental psychology of thought in 1905. His research emphasized the importance of the instructions given to subjects in a specific task; these instructions would create a mental set. Watt was also responsible for the procedure of "fractionation," a modification of experimental introspection in which a complex task would be divided into sections, with introspection then given to each section. While at Glasgow, he published books on the psychology of sound, dreams, and the psychology of music.

Weschler, David (1896–1981). David Weschler, a native of Romania, was a leader in the research area of mental and intelligence testing, best known for creating widely used intelligence tests (e.g., WISC, or Weschler Intelligence Scale for Children; WAIS, or Weschler Adult Intelligence Scale). In 1919, Weschler studied in London with both Karl Pearson and Charles Spearman; during the time of the Miles visit, he was in Paris, studying the physiology of emotion (correlating skin temperature changes with felt emotion) with the physiologist Louis Lapicque and the psychologist Henri Pieron. This work eventually formed the topic of his doctoral dissertation, which he earned from Columbia University in 1925, under the direction of Robert Woodworth.

Whipple, Robert Stewart (1871–1953). R. S. Whipple is best known for his work with the Cambridge Scientific Instrument Company. He studied at King's College in London and worked under his father for eight years at the Kew Observatory before joining the L. P. Casella Firm as a manager. In 1898 he joined the staff of the Cambridge Scientific Instrument Company (CSI). Whipple worked his way up to joint managing director in 1909 and served as chairman of the CSI board of directors from 1939 to 1949. In the early 1920s he started a collection of old instruments and scientific books and in 1944 donated the majority of his collection to the University of Cambridge to form the History of Science Museum and Library. In 1951 the museum and library officially became the Whipple Museum of the History of Science.

Widmark, Erick M. P. (1889–1945). Erik Widmark was a Swedish physician and physiologist who received his MD from the University of Lund in 1917. He accepted a position as an assistant professor of physiology at Lund in 1918 and in 1921 was promoted to professor of medical and physiological chemistry. Widmark is best known for his research on the effects of alcohol consumption. His research on concentration of alcohol in the blood, specifically his micro-method alcohol analysis, led to blood alcohol content (BAC) testing in Swedish motorists as early as 1941. He also studied the influence of food on alcohol absorption in the blood and later studied the behavioral and pathological changes associated with chronic alcoholism in dogs.

Zwaardemaker, Hendrik (1857–1930). Hendrik Zwaardemaker was a Dutch physiologist who was also interested in the science of psychology. In 1897 he was appointed professor of experimental physiology at the University of Utrecht and remained in the position until he retired in 1927. Zwaardemaker is best known for his research in olfaction and his invention of the olfactometer and the term olfactie. He also built an odor-proof room in order to conduct additional research on the minimum perceptible olfaction of humans and animals without external irritation or influence.

References

Agulhon, H., Compton, A., Javillier, M., Macheboeuf, M., Medigreceanu, Mokragnatz, Nakamura, H., Phisalix, C., Rosenblatt, Silberstein, L., & Voronca-Spirit. (n.d.). *Biographical sketch: Gabriel Bertrand (1867–1962)*. Retrieved from http://www.pasteur.fr/infosci/archives/e_bero.html

Andréasson, R. & Jones, A. W. (1995). Erik M. P. Widmark (1889–1945): Swedish pioneer in forensic alcohol toxicology. *Forensic Science International, 72*, 1–14.

Bartlett, F. C. (1923). William Halse Rivers: 1864–1922. *American Journal of Psychology, 34*, 275–277.

Besterman, E., & Creese, R. (1979). Waller—pioneer of electrocardiology. *British Heart Journal, 42*, 61–64.

Bock, B. (n.d.). *John Scott Haldane 1860–1936, Fellow of New College, Oxford*. Retrieved from http://www.giffordlectures.org/Author.asp?AuthorID=73

Boring, E. G. (1965). Edward Wheeler Scripture: 1864–1945. *American Journal of Psychology, 78*, 314–317.

Breathnach, C. S. (1974). Joseph Barcroft (1872–1947), friend and physiologist. *Irish Journal of Medical Science, 143*, 232–237. doi: 10.1007/BF03004768

Brouwer, B. (1923). Professor J. K. A. Wertheim Salomonson Obituary. *Journal of Nervous and Mental Disease, 57*, 322–324.

Bradley, J. K., & Tansey, E. M. (1996). The coming of the electronic age to the Cambridge Physiological Laboratory: E. D. Adrian's valve amplifier in 1921. *Notes and Records of the Royal Society of London, 50*, 217–228.

Bryla, K. (n.d.). *Mellanby, Edward: British physician, 1884–1955*. Retrieved from http://www.faqs.org/nutrition/Biographies/Mellanby-Edward.html

Cohen, S. G. (2002). Pioneers and milestones: Clemens von Pirquet, MD (1874–1929). *The Journal of Allergy and Clinical Immunology: Official Publication of the American Academy of Allergy, 109*, 722–724.

Cunningham, D. J. C. (1964). Claude Gordon Douglas, 1882–1963. *Biographical Memoirs of Fellows of the Royal Society, 10*, 51–74.

Dale, H. H. (1948). Frederick Gowland Hopkins, 1861–1947. *Obituary Notices of Fellows of the Royal Society, 6*, 115–145.

Dekkers, W. J. M. (1995). F. J. J. Buytendijk's concept of an anthropological physiology. *Theoretical Medicine and Bioethics, 16*, 15–39. doi: 10.1007/BF00993786

Denny-Brown, D. (1952). Charles Scott Sherrington: 1857–1952. *American Journal of Psychology, 65*, 474–477.

Donnan, F. G. (1949). Ernst Julius Cohen, 1869–1944. *Obituary Notices of Fellows of the Royal Society, 5*, 666–687.

Douglas, S. R. (1935). George Dreyer, 1873–1934. *Obituary Notices of Fellows of the Royal Society, 1*, 596–576.

Dunn, P. (2000). Sir Joseph Barcroft of Cambridge (1872–1947) and prenatal research. *Archives of Disease in Childhood: Fetal and Neonatal Edition, 82*, 75–81. doi: 10.1136/fn.82.1.F75

Erdman, A. M. (1964). Cornelis Adrianus Pekelharing—A biographical sketch (July 19, 1848–September 18, 1922). *Journal of Nutrition, 83,* 1–9.

Garry, R. C. (1954). Obituary: Edward Provan Cathcart, C.B.E., M.D., D.Sc., LL.D., F.R.S. *British Journal of Industrial Medicine, 11,* 229. doi: 10.1136/oem.11.3.229-a

Geddes, P. (1920). *The life and work of Sir Jagadis C. Bose.* London: Longmans.

Geuter, U. (2000). Felix Kreuger. In A. E. Kazdin (Ed.), *Encyclopedia of psychology.* (Vol. 4, pp. 462–463). Washington, DC: American Psychological Association.

Gottlieb, L. S. (1961). Willem Einthoven, M. D., Ph.D., 1860–1927: Centenary of the father of electrocardiography. *Archives of Internal Medicine, 107,* 447–449.

H., A. (1929). Egerton Charles Grey (1887–1928). *Biochemical Journal, 23,* 1–2.

Harvey, J. (2000). Lapicque, Marcelle (de Heredia) (1873–ca. 1962). In M. Ogilvie and J. Harvey (Eds.), *Biographical Dictionary of Women in Science* (pp. 745–746). New York, NY: Routledge Press.

Hearnshaw, L. S. (1964). *A short history of British psychology, 1840–1940.* London: Butler & Tanner Ltd.

Herbert, S. (n.d.). *Lucien Bull (1876–1972).* Retrieved from: http://www.victorian-cinema.net/bull.htm

Hill, A. V. (1950). August Schack Steenberg Krogh, 1874–1949. *Obituary Notices of Fellows of the Royal Society, 7,* 220–237.

Hill, A. B., & Hill, B. (1968). The life of Sir Leonard Erskine Hill FRS (1862–1962). *Proceedings of the Royal Society of Medicine, 61,* 307–316.

Koehler, P. J., Bruyn, G. W., & Moffie, D. (1998). A century of Dutch neurology. *Clinical Neurology and Neurosurgery, 100,* 241–253. doi: 10.1016/S0303-8467(98)00066-3

Lemmel, B. (n.d.). *Sevran.* Retrieved from http://nobelprize.org/alfred_nobel/industrial/articles/sevran/index.html

Lestel, L. (2007). *Itinéraires de chimistes: 1857–2007, 150 ans de chimie en France avec les présidents de la SFC.* Les Ulis, France: EDP Sciences.

Littman, R. A. (2000). Henri Pieron. In A. E. Kazdin (Ed.), *Encyclopedia of psychology* (Vol. 6, pp. 196–197). Washington, DC: American Psychological Association.

Louis Lapicque, 1866–1952. (1953). *Journal of Neurophysiology, 16,* 97–100.

Lyth, J. C. (1938). Obituary notice. *The British Medical Journal, 1,* 1239.

Magnus, O. (2002). *Rudolf Magnus; Physiologist and Pharmacologist, 1873–1927.* New York, NY: Springer Publishing Company.

Matazarro, J. T. (1981). David Weschler (1896–1981). *American Psychologist, 36,* 1542–1543.

Matthews, D. M. (1978). Otto Cohnheim: The forgotten physiologist. *The British Medical Journal, 2,* 618–619. doi: 10.1136/bmj.2.6137.618

McIvor, A. J. (1987). Employers, the government, and industrial fatigue in Britain, 1890–1918. *British Journal of Industrial Medicine, 44,* 724–732.

Murchison, C. (Ed.). (1929). *The psychological register.* Worcester, MA: Clark University Press.

Myers, C. R. (1982). Psychology at Toronto. In M. J. Wright & C. R. Myers (Eds.), *A history of academic psychology in Canada* (pp. 68–99). Toronto: C. F. Hogrefe.

Neil, E. (1961). Carl Ludwig and his pupils. *Circulation research, 9,* 971–978.

Nobel Foundation (1965). *Physiology or Medicine, 1922–1941*. Amsterdam: Elsevier Publishing Company.

Nobel Foundation (1966). *Chemistry, 1901–1921*. Amsterdam: Elsevier Publishing Company.

Noyons, A. K. M. (1931). Hendrik Zwaardemaker: 1857–1930. *American Journal of Psychology, 43*, 525–526.

Obituary: H. C. Bazett, C.B.E., M.C., M.D., F.R.C.S. (1950). *The British Medical Journal, 2*, 220–222. doi: 10.1136/bmj.2.4672.220

Obituary: C. G. Douglas, C.M.G., M.C., D.M., F.R.S. (1963). *The British Medical Journal, 1*, 890.

Obituary: Prof. Georges Dreyer, C.B.E., F.R.S. (1934). *Nature, 134*, 690–691. doi: 10.1038/134690a0

Obituary: Dr. Egerton Charles Grey. (1928). *The British Medical Journal, 2*, 470.

Obituary: Clemens Pirquet, MD. (1929). *The British Medical Journal, 1*, 526. doi: 10.1136/bmj.1.3558.526

Obituary: Professor Storm Van Leeuwen. (1933). *The British Medical Journal, 2*, 318–319. doi: 10.1136/bmj.2.3788.318b

Oldfield, R. C. (1972). Frederick Charles Bartlett: 1886–1969. *American Journal of Psychology, 85*, 133–140.

O'Connor, W. J. (1991). *British physiologists 1895–1914: A biographical dictionary*. New York: St. Martin's Press.

O'Neil, W. M. (1986). Muscio, Bernard (1887–1926). Australian National University (Eds.), *Australian Dictionary of Biography* (Vol. 10, pp. 650–651). Melbourne: Melbourne University Press.

Pear, T. H. (1947). Charles Samuel Myers: 1873–1946. *American Journal of Psychology, 60*, 289–296.

Porter, T. M. (2004). *Karl Pearson: The scientific life in a statistical age*. Princeton, NJ: Princeton University Press.

Rakemann, F. M. (1958). Professor Willem Storm Van Leeuwen and the asthma problem. *Acta Allergologica, 12*, 407–426. doi: 10.1111/j.1398-9995.1958.tb04009.x

Roughton, F. J. W. (1949). Joseph Barcroft, 1872–1947. *Obituary Notices of Fellows of the Royal Society, 6*, 315–345.

Salomonson, J. K. A. W. (1921). A new ophthalmoscope. *Transactions of the Optical Society, 22*, 53–62. doi: 10.1088/1475-4878/22/2/302

Söderqvist, T. (1996). Partners in physiology [Review of the book *August and Marie Krogh: Lives in science*, by Bodil Schmidt-Nielson). *Science, 271*, 1681–1682.

Someonoff, B. (1951). James Drever: 1873–1950. *American Journal of Psychology, 64*, 283–285.

Spiegelberg, H. (1972). *Phenomenology in Psychology and Psychiatry*. Evanston, IL: Northwestern University Press.

Stephenson, M. (1948). Obituary Notice: Frederick Gowland Hopkins, 1861–1947. *Biochemical Journal, 42*, 161–169.

Tells New Secrets of Plants' Growth (1912, September 7). *The New York Times*, p. 11. Retrieved from ProQuest Historical Newspapers: The New York Times (1851–2006). (104907395).

Thompson, R. A., & Ogston, A. G. (1983). Rudolph Albert Peters. 13 April 1889–29 January 1982. *Biographical memoirs of Fellows of the Royal Society, 29,* 494–523. doi: 10.1098/rsbm.1983.0018

Thorndike, E. L. (1945), Charles Edward Spearman: 1863–1945. *American Journal of Psychology, 58,* 558–560.

The University of Glasgow Story: Biography of Edward Provan Cathcart. (n.d.). Retrieved from http://www.universitystory.gla.ac.uk/biography/?id=WH0035&type=P Thomas, H. H. (1954). Robert Stewart Whipple. *Bulletin of the British Society for the History of Science, 1,* 249–250. doi: 10.1017/S095056360000097X

van Strien, P. J. (2000). Gerardus Heymans. In A. E. Kazdin (Ed.), *Encyclopedia of psychology* (Vol. 4, pp. 125–126). Washington, DC: American Psychological Association.

van Strien, P., J. (2000). Buytendijk, Frederick J. J. (1887–1974). In A. E. Kazdin (Ed.), *Encyclopedia of psychology* (Vol. 1, pp. 492–493). Washington DC: American Psychological Association.

Wagner, R. (1964). Clemens von Pirquet: Discoverer of the concept of allergy. *Bulletin of the New York Academy of Medicine, 40,* 229–235.

Williams, M. E. W. (1994). *The precision makers: A history of the instruments industry Britain and France, 1870–1939.* London, England: Routledge.

Wishart, G. M. (1954). Edward Provan Cathcart, 1887–1954. *Obituary Notices of Fellows of the Royal Society, 9,* 34–53.

Yerkes, R. M. (1942). Raymond Dodge, 1871–1942. *The American Journal of Psychology, 55,* 584–600.

Zwaardemaker, H. (1930). H. Zwaardemaker: University of Utrecht. In C. Murchison (Ed.), *A History of Psychology in Autobiography* (Vol. I, pp. 491–516). Worcester, MA: Clark University Press.

Index

A
Abel, John Jacob, 162
Adrian, Edgar Douglas, 64, 87, 89, 90, 92, 94, 145, 337
alcohol, 15, 36, 37, 53, 55, 76, 101, 104, 108, 109, 125, 150, 165, 209, 241, 254, 257, 258, 266, 299, 301, 302
Alquier, M., 148, 151, 154, 157, 158, 163
American Child Welfare Mission, 292
American Friends Service Committee, 281
Anderson, A. C., 251, 252, 253
Anrep, Gleb V., 11, 12, 13, 14
Arrhenius, Svante, 13, 257, 266, 268
audion amplifiers, 14
audion valves, 11, 14
Aveling, Francis, 16, 17, 51

B
Baird and Tatlock, Ltd., 89
Baldwin, Bird T., 37
Bang, Ivar C., 256
Barcroft, Joseph, 31, 32, 37, 79, 87, 88, 89, 337
Bartlett, Frederick Charles, 88, 96, 97, 98, 337
Bayliss, William Maddock, 9, 10, 12, 14, 43, 101
Bazett, H. C., 105, 107, 109, 110, 111, 229, 338
Bedford College, 66
Benedict, Francis G., 23, 49, 50, 53, 114, 154, 157, 158, 178, 210, 223, 224, 229, 230, 233, 245, 249, 251, 263, 273, 277, 299
Benedict Portable Respiration Apparatus, 73, 158, 198
Bergansius, F. L., 218
Bertrand, Gabriel, 148, 150, 151, 157, 158, 338
bicycle ergometer, 50, 83, 87, 88, 89, 111, 117, 241, 242, 266
Bijtel, J., 218
Binet, Alfred, 149
bio-chemistry, 60, 99, 241, 256, 281
biology, 59, 208
Blix, Magnus G., 261
blood-gas analysis apparatus, 73, 115
Bohr, Christian, 88
Bohr wet meter, 255
Bose, Acharya J. C., 12, 13, 40, 43, 47, 51, 52, 338
Bose Research Institute, 52
Bouckaert, J. P., 177
Boulitte, G., 123, 229
Brielaus (professor), 261
Brinkman, R., 229
British Fatigue Board, 65, 97, 98
British Medical Research Committee, 56, 59, 61, 62
British Physiological Society, 31, 53, 90, 99, 101, 102, 300
British Psychological Association, 66, 300
British Royal Society, 13, 31, 49, 96
Brodie, Thomas Gergor, 101
Brownlee, John, 61, 65
Bruce, A. Ninian, 68
Brugmans, Henri J. F. W., 236, 237, 238
Bull, Lucien, 103, 129, 131, 132, 133, 134, 135, 136, 138, 139, 140, 154, 159, 162, 200, 245, 299, 338
Burrell, L. S. T., 121
Buytendijk, F. J. J., 206, 207, 208, 209, 210, 278, 338–39

C
C. F. Palmer and Company, 60
Cairns, Hugh William Bell, 110
calorimeter, 129, 140, 151, 154, 157, 158, 163, 177, 178, 179, 180, 181, 182, 183, 184, 206, 208, 209, 244, 263
Cambridge and Paul Instrument Co., 103, 203
Cambridge Falling Plate Camera, 14
Cambridge Instrument Co., 21, 27, 28, 111, 249

Cambridge String Carrier, 11, 14
Cambridge String Holder, 14, 28
Cambridge Thread Recording Instrument, 184
camera, 12, 14, 27, 43, 92, 103, 133, 135, 138, 139, 140, 186, 203, 215, 217, 249, 288
Cameron, Charles, 36, 37
Cannon, Walter B., 123
capillary electrometer, 71, 92, 93
Carnegie Institution, 23, 66, 67, 301
Carnegie Nutrition Lab, 21, 24, 30, 31, 38, 39, 55, 58, 61, 68, 76, 78, 82, 84, 97, 101, 103, 105, 108, 110, 117, 123, 124, 136, 140, 150, 154, 157, 158, 163, 165, 184, 198, 208, 209, 212, 224, 238, 255, 261, 262, 263, 266, 273, 277, 280, 281, 294, 299, 301, 307
Carpenter, Thorne Martin, 39, 210, 250, 258, 299, 307
cascograph, 51, 52
Cathcart, Edward Provan, 49, 53, 78, 82, 83, 84, 86, 114, 295, 339
Catholic University of Louvain, 170, 171, 172, 173, 174, 176, 178, 184, 198, 215
chemistry, 13, 30, 59, 143, 170, 212, 215, 216, 256, 266, 267
Chick, Harriette, 295
Child Welfare Institute of Iowa, 37
circuit interrupter, 64
Clark, Roger, 289
Cohen, Ernst, 216, 266, 339
Cohn (a U.S. rep. at Paris Congress), 162
Collot, H., 71
Columbia University, 65, 97, 147
Combe, George, 80
Compton (assistant of Bertrand), 148, 150, 157
"Concertina" apparatus, 73, 74
conditioned reflex, 13
Cornell University, 86
Cotton, Thomas Forrest, 24
Cramer, W., 68
Cremer, Nax, 281
Creveld (assistant of Hamburger), 229
Crommelin, Claude, 227
Cullis, Winifred Clara, 53, 101
Cushny, Arthur Robertson, 73, 76, 78, 165
cyscograph, 12

D
Dale, Henry Hallett, 53, 55, 57, 60, 76, 101, 244
Daly, I. de Burgh, 11, 14, 15
Dansk Maalerfagrik of Copenhagen, 255
Dastre, Jules, 145

Davies, H. W. (Davis), 115, 116
Dawson, P. H., 111
de Decker, G., 42, 44, 164
Dirken, M. N. J., 206, 208
Dodge, Raymond, 250
Donders, F. C., 215
Donegon, Joseph Francis, 11
Douglas, Claude Gordon, 105, 111, 112, 113, 114, 115, 340
Douglas Bag, 111
Drever, James, 80, 340
Dreyer, Georges, 110, 118, 119, 121, 122, 123, 124, 340
Drinker, Cecil Kent, 150
Drummond, Jack Cecil, 9, 13
Duboscq Colorimeter, 111
Dumas, Georges, 149
Durig (professor), 289

E
Edelemann field coils, 14
Edelemann galvanometer, 11
Einthoven, Willem, 28, 160, 186, 217, 218, 219, 222, 273, 340–41
electrocardiogram, 14, 26, 27, 28, 29, 43, 110, 111, 176, 185, 200, 202, 222, 249, 261, 273
electrometer, 92
episcotister, 202, 203, 262,
ergometer, 96, 206, 208, 209, 241, 263, 264, 265
Euclid Computing Machine, 61
Evans, Charles Arthur Lovatt, 101
Exner, Sigmund, 289

F
falling plate camera, 27, 203
Fano, G., 162
Fauville (student), 177
Feil, H. S., 28, 29
Ferings (assistant of Hamburger), 229
Field, Herbert Haviland, 165
Flack, Martin William, 58
Fleming and Marconi valves, 14
Fletcher, Walter, 32, 62, 65, 96
Flexner, Simon, 162
Flugel, J. C., 16
Forbes, Alexander, 14
Francis Galton Eugenics Laboratory, 22, 23, 61
Free University, 208

G
Galeotti, G., 299

galvanometer, 14, 21, 24, 27, 28, 29, 47, 48,
    71, 83, 93, 95, 103, 129, 133, 134, 136, 138,
    139, 176, 177, 184, 186, 200, 202, 203, 204,
    205, 217, 249, 287, 288
Garten, Siegfried, 287, 288
gas analysis apparatus, 245, 247, 249
gas-mixing meter, 119
General Electric Company, 124, 135
Gildemeister, Martin, 279, 280, 281
Gjellerup, J., 255
Gley, Eugène, 146, 162
Greenwood, Major, 55
Gregory, R. A., 65
Grey, Egerton Charles, 99, 100, 341
Gryns (researcher), 213, 215
Guy's Hospital Medical School, 33, 35, 38, 39

H
Haldane, Jack, 115, 116
Haldane, John S., 73, 74, 88, 113, 115, 116,
    117, 245, 341
Halliburton, W. D., 18, 30, 31, 101, 341
Hamburger, H. J., 86, 160, 171, 186, 229, 230,
    231, 233, 234, 236
Hamilton, William, 15
Hanson, George F., 119, 120, 122, 123
Hardy, W. B., 65
harmonic analyzer, 61, 65
Harris, J. Arthur, 23
Hartree, William, 87, 94, 95
Harvard Apparatus Company, 73
Harvard Medical School, 14, 111, 123, 169, 217
Harvard University, 125, 150
Harwood-Ash, Miss D., 56, 57
Haycraft, J. B., 101
Heger, Paul, 17, 169, 170, 279
Henderson, Lawrence J., 162
Henriques, V., 249
Hering, Karl E. K., 287, 288
Hess, Alfred F., 162, 289
Heymans, Gerard, 233, 236, 237, 342
Higgins, Harold Leonard, 89, 299
Hill, Archibald Vivian, 94, 95, 101
Hill, Leonard, 55, 56, 57, 58, 59, 65, 82, 244, 342
Hill of Lund, 249, 273
Hindhede, M., 249
histology, 9, 30, 59, 68
Hobson, Frederick Greig, 121
Holima Autofrigor, 209
Hopkins, Frederick Gowland, 55, 99, 342
Hughes, E. C., 37

Humberstone, T. L., 51
Hunt, Reid, 224

I
Imperial Cancer Research Commission, 68
Industrial Inst. for Applied Psych. in England, 32
Institute for Preventive Medicine, 65

J
James Hicks Company, 56
Janet, Pierre, 149
Johansson, Johan E., 160, 263, 264, 265, 266,
    267, 268, 273, 274, 342
Jones, L. Wynn, 16, 19, 21
Jung, Carl, 125

K
Kapteyn, J. C., 234, 235
Karolinska Institute of Stockholm, 263, 264
Karolinska University of Lund, 256, 261
"kata"-thermometer, 56, 57
Keogh, Alfred, 65
Kestner, Otto (Cohnheim), 160, 164, 226, 277,
    278, 343
King's College for Women, 34, 53
King's College Medical School, London, 18,
    30, 31
Kirschmann, August, 284, 286, 343
Klas, Sondén, 267, 268
Kline, Major, 66, 67
Krogh, August, 49, 210, 222, 241, 242, 243,
    244, 245, 246, 247, 248, 250, 343
Krogh, Marie, 241
Krueger, Felix, 284, 343–44
Kulpe, Oswald, 85
kymograph, 28, 51, 60, 73, 83, 105, 108, 129,
    203, 243, 247–48, 261, 262, 288

L
Laidlaw, Patrick Playfair, 37, 100
Langlois, Jean Paul, 144
Lapicque, Louis, 142, 143, 145, 158, 344
Lapicque, Marcelle, 142
Le Goff, Jean, 165
Lee, Frederic Shiller, 65, 97, 162, 164
Lefevre, Jules, 148, 154, 157, 158, 163
Lehmann, Alfred, 250
Lemoine (friend of Lefevre), 148, 157
Levine, S. A., 24
Lewis, Thomas, 24, 26, 27, 28, 29, 273
Libbrecht (instructor), 172, 177

Liljestrand, Goran, 218, 222, 223, 273
Lim, R. K. S., 68
Lindhard, Johannes L., 249
Lister Institute, 59, 88, 99, 295
Lodge, Oliver, 12
London University, 40, 41
Lowell Institute, 171
low-pressure chamber, 120, 123–24
Lucas comparator, 28, 29
Lucas pendulum circuit breaker, 92, 145
Lucas, Keith, 64, 89, 90, 92, 94
Ludwig, Carl, 287
Lusk, Graham, 162, 165

M

Magnus, Rudolf, 222
Malengreau, Fernand, 177
Mallenby, E., 34, 53, 54, 101, 295, 344
manometer, 60, 83, 210
Marbe, Karl, 85
Marey Institute, 103, 129, 131, 132, 134, 136, 140, 141, 299
Martin, C. J., 88
Massachusetts Institute of Technology, 94
Maudsley Hospital, 125
McCall, J., 83
McDougall, William, 55, 61, 80, 81, 125
McKenzie, James, 65
Meakins, J. C., 73, 74, 75, 110, 115
Meeting of the Physiological Society in Cambridge, 31
metabolism, 13, 14, 31, 33, 53, 58, 82, 118, 123, 244, 250
Michaelin, L., 254
micro-gas analysis apparatus, 210
Miles, Walter R., 41, 107, 144, 268, 307
Military Hospital of Stockholm, 273
Military Medical Academy, Russia, 13
Mills, E. S., 300
Ministry of Health, London, 55
Mollgaard, Holger, 251, 252, 253
Moore, Benjamin, 32, 58, 59, 60
motion picture camera, 132, 140
motion pictures, 140, 141, 164, 230
Muller-Lyer Illusion, 236
Murschhauser, Hans, 299
Muscio, Bernard, 96, 97, 98, 344
Museum of Natural History, Paris, 145
Myers, Charles S., 32, 65, 96, 98, 344–45
myograph, 105, 107

N

National Institute for Applied Psychology and Physiology, 65, 96, 98
National Institute for Medical Research, 56, 57, 63, 65, 76
National Institute for Nutrition in Holland, 193, 301
National Physical Laboratory, 65
Natural History Dept., British Museum, 32
Netherlands Institute for Nutrition, 193, 198, 199
Newman, George, 55
Nicloux, Maurice, 165
Nogues, Pierre, 140, 141
Noyons, A. K. M., 163, 170, 171, 172, 174, 177, 178, 179, 180, 182, 183, 184, 185, 186
Nutrition, 9, 36, 38, 60, 142, 145, 193, 198, 199, 279, 289, 292, 293, 294, 295, 301

O

olfactometer, 212
Olsen (researcher), 250
Onnes, H. Kamerlingh, 227
opthalmoscope, 200
oxygen, 37, 58, 73, 87, 114, 117, 227

P

Pares, Bernard, 31
Pasteur Institute, 99, 150–51
pathology, 118, 223
Paton, D. Noel, 82
Paulta (professor), 295
Pavlov, Ivan, 11, 12, 13, 14, 31, 278
Pearson, Karl, 21–22, 23, 61, 345
Pekelharing, Cornelis Adrianus, 212, 214, 215, 345
Pembrey, Marcus Seymour, 12, 33, 34, 35, 36, 37, 38, 345–46
Pennsylvania Hospital in Philadelphia, 28
Perry, E. Cooper, 65
Peters, Rudolph Albert, 99, 346
pharmacology, 60, 76, 78, 193, 224
Phillipson (professor), 169, 170
Philprit, Captain, 16, 17, 20, 21
phrenology, 80
physics, 59, 227, 261, 286
Physiological Congress, 31, 34, 53, 141, 144, 145, 146, 160, 162, 163, 164, 177, 222, 229, 233, 234, 245, 279, 287, 295, 299, 300
physiology, 9, 10, 12, 13, 30, 33, 36, 39, 40, 49, 53, 55, 64, 65, 68, 71, 72, 73, 82, 83, 85, 86, 87, 92, 94, 96, 98, 106, 109, 118, 142, 143, 160, 162, 169, 170, 171, 174, 177, 185, 208,

212, 215, 217, 222, 223, 226, 229, 230, 233, 234, 241, 248, 251, 256, 261, 266, 273, 277, 279, 280, 281, 284, 287, 295, 299, 300, 301
Pierce, H. F., 120, 123, 124
Pieron, Henri, 147, 148, 149, 346
Plimmer, Aders, 9
pneumograph, 147, 224
Porter, William Townsend, 111
Poulton, Edward Palmer, 37
Priestly, J. G., 73, 74
psychiatry, 32, 96
psychoanalysis, 125
Psychoerogograph, 21
psycho-galvanic reflex, 21, 47, 48, 80, 96, 147
psychology, 16, 32, 61, 65, 80, 81, 82, 83, 85, 86, 88, 94, 96, 98, 109, 125, 147, 149, 210, 212, 236, 257, 284
pursuit meter, 15, 19, 97, 144, 163
pursuit pendulum, 15, 97, 98, 101

R

Read, Cavenish, 67
respiration, 49, 73, 74, 83, 111, 115, 123, 147, 157, 158, 179, 180, 198, 209, 223, 248, 251, 253, 255, 263, 264, 273, 277
respiration chamber, 79, 111, 113, 114, 115, 124, 172, 177, 185, 241, 242, 253, 263, 265
Richet, Charles, 144, 162, 164
Ringer, W. E., 212, 214
Ritzman, E. G., 251
Rivers, W. H. R., 96, 101, 346
Rivers-McDougall Dotting Machine, 61, 96
Rubner, Max, 86, 160, 162, 164, 279, 280
Russell, William Thomas, 61
Ryffel, John Henry, 33, 39

S

St. John's College, 101
Salomonson, J. K. A. Wertheim, 186, 200, 201, 202, 203, 204, 205, 251, 347
Schafer, Edward Albert Sharpey, 55, 68, 72, 83, 165
Schick, Béla, 289
Schuster, Edgar, 61, 62, 63, 65
Scientific Academy of the Nobel Institute, 266
Scripture, E. W., 66, 67, 347
Seashore, Carl, 20, 66
Secker, H., 79
Sedgwick, W. T., 94
Sherrington, Charles, 55, 65, 97, 105, 106, 107, 108, 109, 164, 347
Siebe-Gorman Company, 73

signal galvanometer, 203
slide rule, 119, 123
slow moving coil galvanometer, 47
Smith, H. M., 299, 300
Smith, W. G., 80
societe scientifique d'hygiene alimentaire, 148, 154, 159
Solvay Institute, 169
Sorensen, W., 249
sound proof room, 16, 96, 215, 236
Souza, G. H. de Paula, 307
Spearman, Charles S., 16, 17, 19, 21, 22, 66, 286, 347–48
spectroscope, 288
Spindler and Hoyer, 80
spirometer, 243, 245, 246, 247, 248
Starling, Ernest, 9, 11, 55, 65, 230
Stass, M., 169, 170
State University of Groningen, 171, 229, 230, 234, 236
State University of Leiden, 217, 223, 224, 227
State University of Utrecht, 171, 212, 216, 222, 266
Stenström, N., 268, 273, 274
Stephenson, Marjory (Stevenson), 99
Stewart, G. N., 162
Stockholm Board of Health, 267
Stodel (associate of Lapicque), 142, 143, 145
string galvanometer, 11, 14, 24, 29, 103, 129, 131, 133, 134, 136, 139, 185, 186, 200, 203, 204, 205, 217, 222, 287, 288
Strömbeck, J. P., 268
Stroud, W. D., 28, 29
Sturgess, T. N., 89

T

Tait, John, 68
Talbot, Fritz B., 255
Tangl, Francis, 217, 223
Taylor, Alonzo, 60
Taylor, W. W., 68, 71, 75
Thane, George, 12, 34, 38, 348
thermocouple, 103
thermopiles, 94–95
Thomas, Karl, 279, 280
thread recording galvanometer, 111
Thunberg, Torsetn, 256, 261
Tigerstedt, Carl, 299
Tigerstedt, Robert A., 160
Titchener, Edward Bradford, 86
Towne, Robert, 37, 38

treadle ergograph, 83
Trinity College, 94, 99
Tucker, William S., 134
tuning fork, 27, 139, 203

U

University College Hospital, 24
University College, London, 9, 10, 12, 16, 17, 22, 23, 43, 61, 76,
University of Amsterdam, 185, 198, 200, 201, 262
University of Berlin, 279, 281
University of Bristol, 89
University of Brussels, 17, 169
University of Cambridge, 31, 32, 37, 87, 92, 96, 98, 99
University of Copenhagen, 241, 242, 249, 250
University of Edinburgh, 68, 73, 74, 75, 76, 78, 80, 96, 115
University of Glasgow, 78, 82, 85
University of Hamburg, 277
University of Hungary, 223
University of Leeds, 16
University of Leiden, 193
University of Leipzig, 284, 287
University of London, 58
University of London Club, 51
University of Oxford, 61, 81, 105, 106, 118, 119, 124
University of Paris / the Sorbonne, 142, 143, 145, 147, 148, 164
University of Strasbourg, 165
University of Toronto, 284
University of Vienna, 289, 295

V

van der Heyde, Henri Christiaan, 210
van Leersum, E. C., 193, 195, 198, 208, 209, 210, 212, 213, 214, 215, 216
van Leeuwen, Willem Storm, 160, 222, 223, 224, 225, 226, 227, 278, 348
van 't Hoff, J. H., 216
Verdin dry meter, 123
Verzar, Frigyes, 217, 218, 223
Veterinary High School of Copenhagen, 251
vitamins, 31, 54, 99
von Pirquet, Clemens, 60, 289, 292, 293, 294, 348

W

Walker, May, 68
Waller, Alice, 40, 41, 42, 43, 44
Waller, Augustus, 21, 31, 40, 41, 43, 44, 45, 47, 48, 49, 50, 164, 348–49
Waller, Jack, 40, 43, 52
Wasserman Commission, 62
Watt, H. J., 85, 86, 349
Weber, E., 281
Wechsler, David, 147, 349
Weinberg (assistant of Hamburger), 229
Whipple, Robert Stewart, 103, 349
Widmark, Erik M. P., 241, 254, 256, 257, 258, 261, 350
Williams bottle, 277
Wilson, D. R., 97
Wirth, Wilhelm, 286
Woodhead, G. S., 103
Woodworth, Robert S., 147
Wundt, Wilhelm, 85, 284
WWI, 9, 14, 19, 24, 29, 39, 55, 68, 82, 85, 86, 96, 111, 123, 125, 129, 134, 139, 142, 154, 158, 159, 160, 169, 170, 171, 174, 177, 209, 256, 263, 267, 277, 278, 280, 281, 284, 287, 299, 301

X

X-ray equipment, 200

Ziss Refractometer, 184
Zuntz, Nathan, 281
Zunz, E., 170
Zwaardemaker, Hendrik, 171, 185, 212, 213, 215, 350